THE IMPACT OF THE FEDERAL BUDGET PROCESS ON NATIONAL FOREST PLANNING

Recent Titles in
Contributions in Economics and Economic History

Corporate Hegemony
William M. Dugger

Taiwan's Development: Implications for Contending Political Economy Paradigms
Cal Clark

Income Inequality in the United States, 1947-1985
Nan L. Maxwell

Evolution of United States Budgeting: Changing Fiscal and Financial Concepts
Annette E. Meyer

Charitable Giving and Government Policy: An Economic Analysis
Jerald Schiff

Economic Law and Economic Growth: Antitrust, Regulation, and the American Growth System
George E. Garvey and Gerald J. Garvey

The Politics of Industrial Recruitment: Japanese Automobile Investment and Economic Development in the American States
Ernest J. Yanarella and William C. Green, editors

The Theory of Contestable Markets: Applications to Regulatory and Antitrust Problems in the Rail Industry
William B. Tye

Advertising and the Transformation of American Society, 1865-1920
James D. Norris

Privatization in the Developing World
L. Gray Cowan

Nuclear Power Goes On-Line: A History of Shippingport
William Beaver

THE IMPACT OF THE FEDERAL BUDGET PROCESS ON NATIONAL FOREST PLANNING

V. ALARIC SAMPLE

Foreword by Dennis C. Le Master

Contributions in Economics and Economic History, Number 103

Greenwood Press
New York • Westport, Connecticut • London

Library of Congress Cataloging-in-Publication Data

Sample, V. Alaric.
The impact of the federal budget process on national forest planning / V. Alaric Sample ; foreword by Dennis C. Le Master.
p. cm.—(Contributions in economics and economic history, ISSN 0084-9235 ; no. 103)
Includes bibliographical references.
ISBN 0-313-27202-6 (lib. bdg. : alk. paper)
1. United States. Forest Service—Appropriations and expenditures. 2. Forest policy—United States. 3. Forest reserves—United States—Planning. I. Title. II. Series.
SD565.S27 1990
333.75′15′0973—dc20 89-27282

British Library Cataloguing in Publication Data is available.

Library of Congress Catalog Card Number: 89-27282
ISBN: 0-313-27202-6
ISSN: 0084-9235

First published in 1990

Greenwood Press, 88 Post Road West, Westport, CT 06881
An imprint of Greenwood Publishing Group, Inc.

Printed in the United States of America

The paper used in this book complies with the Permanent Paper Standard issued by the National Information Standards Organization (Z39.48-1984).

10 9 8 7 6 5 4 3 2 1

To Dory, for her patience and cheerful encouragement,

and my bookends, Rebecca and Emily.

Contents

Tables and Figures

Tables

Figures

Foreword

One of the reasons V. Alaric Sample's study on budgeting and national forest planning is important is because of its unique focus on policy implementation. In the literature of public administration, much has been written on the formulation of public policy. Very little has been offered on the far more complicated and uncertain process of implementing public policy. This book helps fill that gap.

No two processes have more day-to-day impact on public policy than the preparation of the president's annual budget request in the executive branch and annual appropriations in the legislative branch. Passing legislation and adjudicating lawsuits are more dramatic, familiar and accessible to the public. Hence, they are often written about. But passing legislation and adjudicating lawsuits tend to be discrete events, rather than continuous activities. Thus, they offer fewer opportunities for affecting policy.

In contrast, preparation of the president's budget request is virtually a year-round activity, largely screened from public view, in which important priorities are set. Similarly, the appropriations process also continues throughout the year, setting priorities through funding levels and making policy through the inclusion of special language in the appropriations acts and accompanying committee reports. While the public can participate in the appropriations process by taking part in hearings and lobbying, the process is more complicated than the authorizing process and requires a level of knowledge and sophistication that few citizens have.

The Office of Management and Budget and the Office of the Secretary of Agriculture play leading roles in the preparation of the president's budget proposal for the Forest Service. Within Congress, most of the key decisions are made within the appropriations committees by subcommittees such as the House Appropriations Subcommittee on Interior and Related Agencies. The procedures within these institutions, and the values, interests and personalities of the individuals occupying key positions in them all affect how policy will be implemented.

The National Forest Management Act mandated reform in the managment of the national forests so timber harvesting could continue. But public policy is no more than a statement of intent until the necessary funds have been requested and appropriated. Sample's study is the first to explore and document the respective roles of the Forest Service, Office of the Secretary of Agriculture, Office of Management and Budget, and the congressional appropriations committees in determining whether national forest management will reflect the reforms mandated by the National Forest Management Act.

The struggle to integrate the budget process with strategic and program planning is a fundamental problem in administration. It is a primary concern and occupation of executives and managers in nearly any large organization, public or private. Furthermore, public administrators confront the additional challenge of not having the discretion or authority to match spending priorities with the priorities established through agency planning. They must yield this authority to other executive and legislative leaders, who often have conflicting views over what the near-term priorities should be.

This book presents a thorough exploration of the failure to integrate budgeting and planning as it affects the implementation of the National Forest Management Act. The principles, however, are generalizable to the implementation of many laws and public policies. Agency heads will find Sample's analysis insightful and his recommendations thoughtful and practical, and perhaps provocative at times. This book is a significant and welcome contribution to the small but growing body of literature on public policy implementation.

Dennis C. Le Master
Professor and Head
Department of Forestry
and Natural Resources
Purdue University

Preface

A public policy is no more than a statement of intent until, and unless, it is successfully implemented through often complex administrative processes. Chief among these is budgeting, a resource allocation process through which policies can be either promoted or suppressed. In the management of the national forests, as in many areas of federal policymaking, the budget precess includes an array of organizations and suborganizations in both the executive and legislative branches, each with their own values, incentives, and agendas. The interplay of these powerful forces, and their impact on the resources of our national forests, is the subject of this study.

In response to continuing controversy over the management of the national forests, Congress gave the U.S. Forest Service two powerful and far-reaching mandates for developing comprehensive, long-range resource management plans, with extensive public involvement, for the national forests. Agency-wide strategic planning was mandated by the Forest and Rangeland Renewable Resources Planning Act (RPA) of 1974. Processes for developing management plans for each national forest were prescribed by the National Forest Management Act (NFMA) of 1976. The NFMA also instituted a number of reforms covering specific management practices, such as clearcutting, to govern the activities of the Forest Service. Several strategic plans have been issued under RPA over the past decade, and management plans have now been completed for most of the 156 national forests.

But will these plans be effective in guiding the actual resource management activities that take place "on the ground" in each national forest? Will the reforms and other legislative requirements of the planning mandates truly make a difference in how the national forests are managed? A plan is little more than a collection of paper until priorities are established, adequate funding is provided, and there is the will among local managers to undertake the tasks called for. Both program planning and budgeting are critical processes in managing and directing any large organization, especially one as decentralized as the Forest Service. To work effectively, the two processes must be well integrated; they must be supportive of and complementary to one another, not pulling the organization in different directions.

Integration of the results of program planning at the field level with the budget process is attainable, but there are substantial barriers to that integration. Some of these barriers arise in the agency's external environment, in the Congress or elsewhere in the executive branch with the Department of Agriculture or the Office of Management and Budget (OMB). Considerable barriers can arise because of the inherently different conceptual approaches used in planning from those used in developing annual budgets.

This is a study of the difficulties that have occurred in integrating the results of strategic planning under RPA with the annual budgets for the Forest Service, and the difficulties that can be expected to occur in implementing the national forest plans prepared under NFMA if corrective measures are not taken. As a basis for determining what those corrective measures are and how successful they might be, the external political environment of the Forest Service, as well as the agency's internal dynamics, are explored in discussed in detail. How are local management plans used in setting priorities during the agency's internal budget development process? How are Forest Service budget proposals modified by the Secretary of Agriculture, OMB, and Congress, and what are the values and incentives that motivate them to make those changes? To what degree do the changes made to Forest Service budget proposals limit the abilities of local managers to carry out the management activities depicted in their program plans? To what degree does the flexibility in the agency's accounting and reporting procedures limit Congress' ability to exercise its oversight responsibilities over the Forest Service?

These and other questions are addressed in the pages that follow. The study begins with a critical examination of budget trends for the Forest Service over the past 25 years. On a program-by-program basis, each year's budget is analyzed at each point in the political process to reveal the changes made by the Secretary of Agriculture, OMB, and the House and Senate appropriations committees. The results are displayed against Forest Service strategic plans developed under RPA to determine the consistency between planned and actual budgets and program levels. To get a better sense of the internal dynamics and external environment of Forest Service budget development,the budget for a single fiscal year is followed from its initial stages on five case study national forests to its final approval as Forest Service appropriations. The budget is then followed through two additional and often overlooked stages: the allocation of funds back to the local national forest level, and the reporting of expenditures at the end of the fiscal year. Such a sequence permits the comparison, on a program-by-program basis, of the funding that was (1) requested, (2) allocated, and (3) reported spent. Differences between the various stages are analyzed, and there is a discussion of the possible reasons for these discrepancies.

The lack of integration between Forest Service planning and budgeting processes has important implications for the implementation of national forest plans and, more broadly, for the carrying out of the various reforms of the National Forest Management Act. This raises larger questions that are of concern not only to those directly interested in the management of the national forests, but also to those concerned more broadly with the effectiveness of public policymaking, and with the functioning and responsiveness of large, public bureaucracies in a democratic society. This book is thus intended for both those readers interested in the development and implementation of national forest policy, and for those concerned more generally with the administration of large, decentralized public agencies. Through its exploration of the motives and power of the institutions involved in budget development for the Forest Service, it offers insights for greater influence as well as greater understanding, However, practitioners and students of public management and policymaking will find this study to be a clear and thorough illustration of general principles, developed in the context of current administrative theory, that apply to the implementation of public policies at all levels.

Acknowledgments

Although this study has been a very personal undertaking, it is the result of the efforts, advice, assistance and support of literally dozens of people. The many individuals who played a role in facilitating the months of research are too many to name here, but the efforts of each and every one are deeply appreciated. Substantial intellectual contributions were made by James W. Fesler, Professor Emeritus of Government in the Department of Political Science at Yale University, Clark S. Binkley, Professor of Forest Economics at Yale, and Garry D. Brewer, Professor of Forestry and Organization and Management at Yale. I also wish to thank Albert C. Worrell, Professor Emeritus of Forest Policy and Economics at Yale, for his contributions and for his encouragement when it was most important.

The research would have been nearly impossible without the cooperation and assistance of the land management planning and policy analysis staffs of the U.S. Forest Service. For their foresight, confidence and support, I extend a special thanks to Jeff M. Sirmon, Deputy Chief for Programs and Legislation, and to Everett L. Towle, Director for Land Management Planning. Much of the remainder of the research was conducted as a senior fellow at the Conservation Foundation in Washington, D.C.. Senior associate William E. Shands has been an unflagging source of encouragement, editorial advice, and thoughtful analysis throughout.

Major technical support was provided by the Program Development and Budget staff at the Washington office of the Forest Service; a special thanks to John A. Leasure, Robert E. Gordon and Marilyn Chilton for their advice and patience. A debt of gratitude is also owed to Michael MacDonald in the Department of Agriculture Office of Budget and Program Analysis, Steven P. Satterfield formerly of the Office of Management and Budget, Kathleen R. Johnson of the House Interior Appropriations Subcommittee staff, and Donald R. Knowles of the Senate Interior Appropriations staff. A number of reviewers spent a great deal of time and effort to help me avoid straying too far afield. Of particular help in this regard were Robert E. Wolf, formerly of the Congressional Research Service at the Library of Congress, Roger A. Sedjo of Resources for the Future, Dennis C.

Le Master of Purdue University, Carl Reidel of the University of Vermont, Lloyd Irland of Maine's Irland Group, and Herbert Kaufman of Yale University.

A special debt of gratitude is owed to Arnold W. Bolle, Dean Emeritus of the University of Montana School of Forestry, for his continuing inspiration and guidance, and for his unwavering faith and encouragement from the very beginning of this endeavor through to its completion.

This effort would not have been possible without the enthusiastic and steadfast support of Jennifer and Ted Stanley, whose efforts on behalf of our national forests will benefit us all for decades to come. Major support was also provided by the Jonathan Sachs Foundation, the Laird, Norton Foundation, and the David and Lucile Packard Foundation. The National Wildlife Federation provided substantial support through its 1987 Environmental Conservation Fellowship Award for Excellence in Graduate Research.

Though information and advice for this study was provided by many, the interpretation and presentation remains the sole responsibility of the author. The events described herein were observed from many different perspectives, all of which were carefully considered in developing the interpretation of those events. To paraphrase Professor Kaufman, I have followed the facts wherever they seemed to me to lead, even when I was thereby compelled to modify or reject suggestions. It is my hope that this perspective, being the product of many, might reveal more about the forces that motivate national forest planning and policymaking than any one perspective taken alone.

THE IMPACT OF THE FEDERAL BUDGET PROCESS ON NATIONAL FOREST PLANNING

1

Introduction to National Forest Planning

The National Forest Management Act (NFMA) of 1976 brought about major reforms in the management and protection of the lands and resources of the National Forest System (NFS). This was an era of sharp public debate over the Forest Service's management of the national forests, brought to a crescendo by a lawsuit that halted all national forest timber harvesting in much of the East. As the principal author of legislation designed to help resolve the controversies, Senator Hubert Humphrey stated:

> We are so constantly embroiled in issues of immediate focus that we are failing to look to the long term . . . we work too much on an ad hoc basis in the Congress and the Executive Branch, moving from crisis to crisis, applying policy and funding Band-Aids and aspirin to long-term problems that require permanent treatment. This is particularly true of our forests and rangelands.[1]

In introducing both the NFMA and, earlier, the Forest and Rangelands Renewable Resources Planning Act (RPA), Humphrey was expressing his conviction that the resources of the national forests should be managed not on the basis of short-sighted annual budget decisions, but on the basis of publicly affirmed long-term goals and objectives. Over the past decade, the Forest Service has been developing long range resource management plans for the 156 national forests, as directed by RPA and NFMA. The importance of these plans to the future of the 191 million acre National Forest System is evident in the substantial investments of time and resources being made by the Forest Service to develop the plans, and by the forest products industry, conservation organizations and individual citizens to monitor and participate effectively in the decision process.

Much of this effort to produce management plans that will be useful and effective guidelines for operational decision making by Forest Service field officials, and that are generally accepted by the multitude of individuals and organizations with an interest in national forest management, could ultimately have little impact on actual resource management activities. Which resource programs are carried out as specified in the forest plans, and which programs will continue along their current course *despite* what is called for in the plans, will actually be determined in the budget development and Congressional appropriations process-a process that few national forest users fully understand or can hope to have effective influence over. One of the greatest concerns facing local national forest managers themselves is that the current budget and appropriations processes will derail the years of effort that have gone into national forest planning. The Forest Service has struggled for many years to more effectively integrate its program planning processes with the budget development process, but with only limited success. If resource planning on the national forests is to be useful, however, such an integration must be achieved. At stake is whether or not the management of the nation's forest resources will be on the basis of long-term goals, as envisioned by Humphrey, or will continue to be driven by political decisions that seldom look farther ahead than next year's budget.

The Forest Service and the National Forest System

The Forest Service, established in 1905 as an agency in the U.S. Department of Agriculture (USDA), is responsible for the protection and management of the 191 million acre National Forest System (NFS). With approximately 35,000 full-time employees it is by far the largest agency in the USDA, yet it accounts for only about seven percent of the USDA annual budget.[2] Although the Forest Service is also responsible for programs in forest research and state and private forestry assistance, operation of the National Forest System is the agency's largest activity and, when road construction is included, accounts for nearly three-quarters of its annual budget of approximately $2 billion.[3] Line management of the NFS is decentralized from the Washington Office to nine regional offices, under which are grouped the 156 national forests (see Figure 1.1). The area of a national forest ranges from 400,000 to 3,000,000 acres and is comprised of from four to nine ranger districts.[4] Line officers--the chief, regional foresters, forest supervisors and district rangers--are responsible for the development of programs for their units and the management of allocated funds (see Figure 1.2). Each line officer is supported by a staff of specialists in such resource programs as timber, range, recreation, and fish and wildlife (see Figure 1.3). However, strong informal ties exist between the functional specialists at the different administrative levels in the organization, and it is common for specialists in any given area to communicate directly on both technical and management issues.

Management of the national forests was largely custodial until the mid-1950s, when the nation's demand for forest resources, primarily

Figure 1.1
The National Forest System and Location of the Five Case Study National Forests

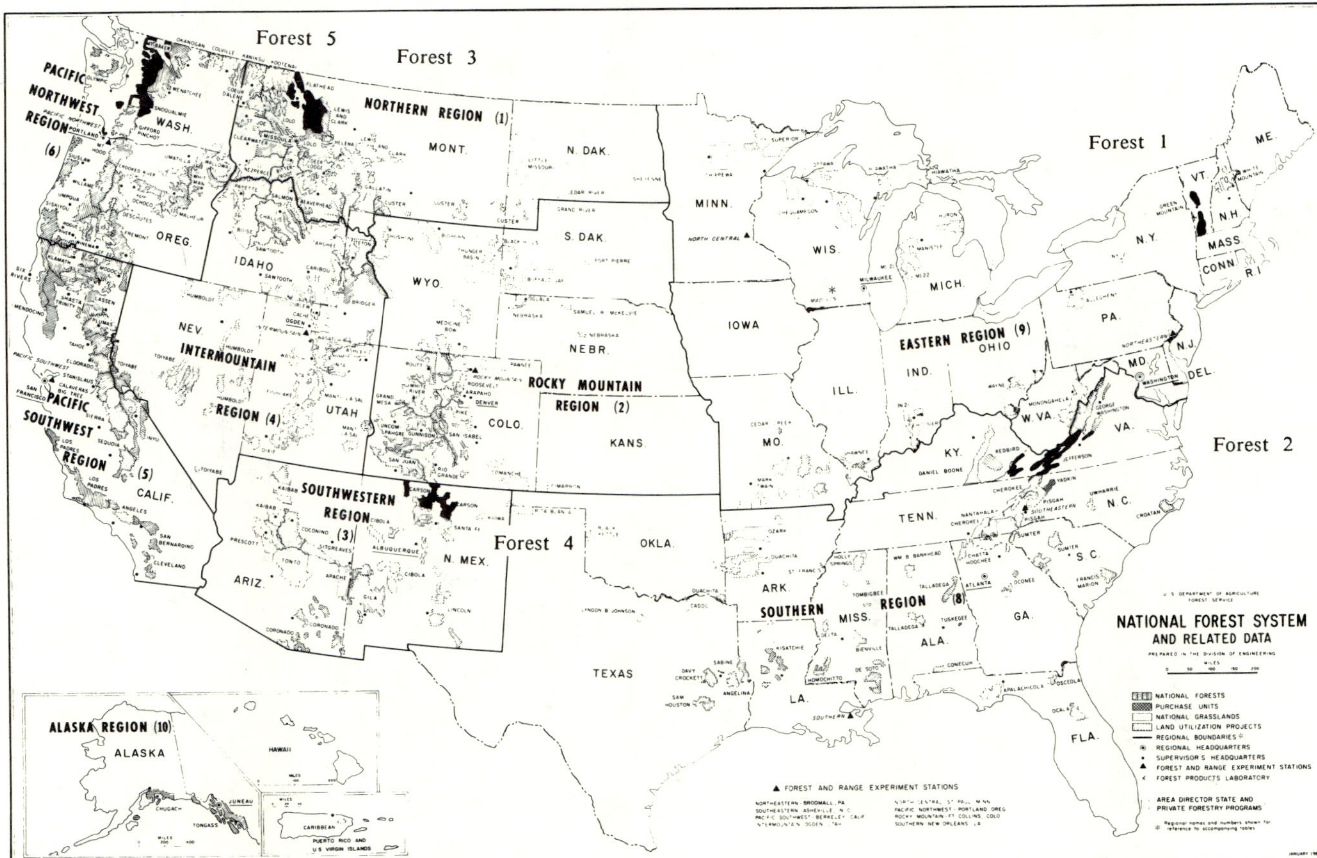

Figure 1.2
Organization of the USDA Forest Service

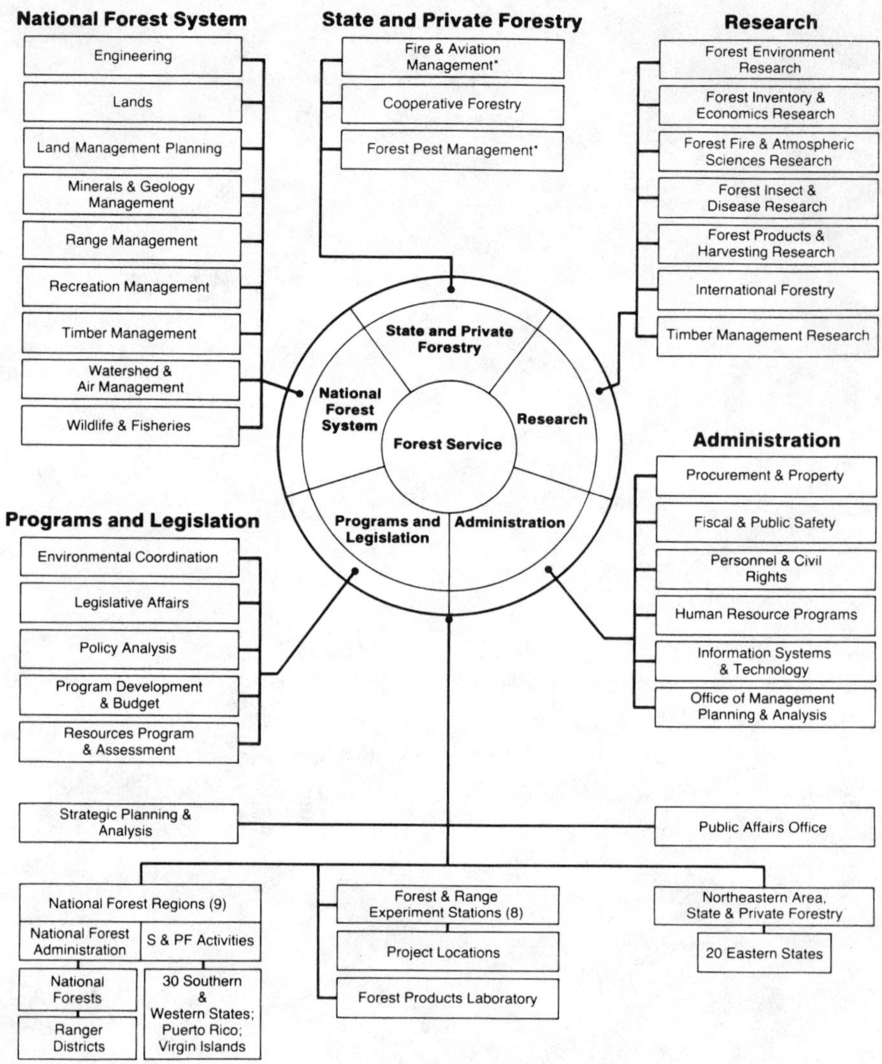

Figure 1.3
Organization of the National Forest System

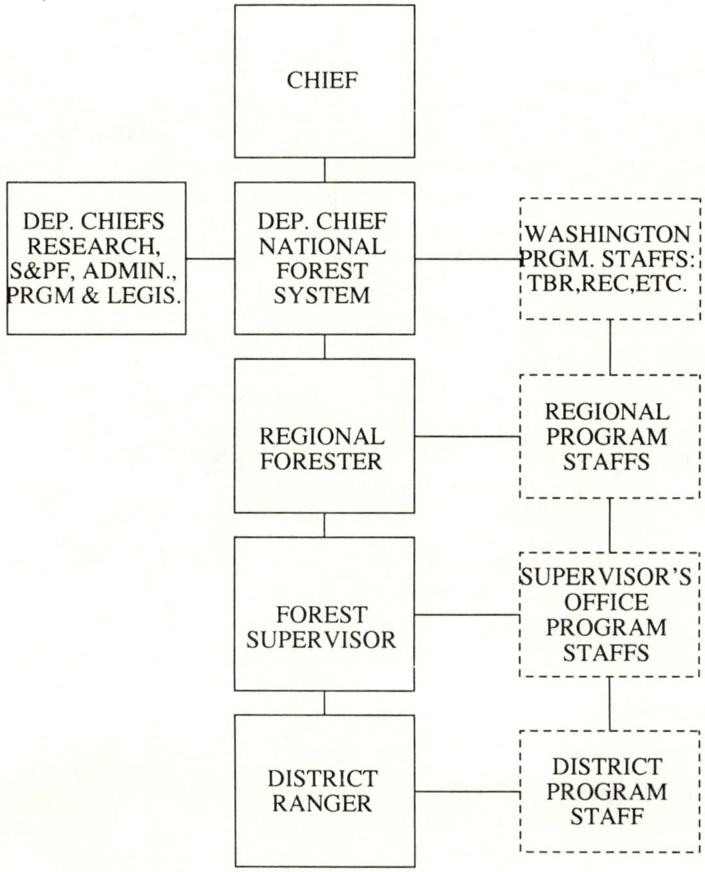

timber and outdoor recreation, began to escalate rapidly. The Multiple Use-Sustained Yield Act of 1960 established that the national forests were to be managed for a variety of resources including timber, range, wildlife, recreation and water. Wilderness was included among the multiple uses with the passage of the Wilderness Act in 1964. As demand for each of these resources continued to grow, the inevitable conflicts arose, and striking a workable balance among competing resource uses became one of the most important and challenging duties of Forest Service field managers. In the early 1970s, public disagreements with the Forest Service over the balance between commodity uses, such as timber harvesting, and non-commodity uses, such as recreation, boiled over into the courts. Ostensibly over timber management practices such as clearcutting, these battles raised serious questions regarding the quality of land management on the national forests and the responsiveness of the Forest Service to public demands for greater emphasis on non-commodity resources. Ultimately, they set the stage for legislative action by the Congress with the objective of getting the Forest Service out of the courts and providing clearer direction for the management of the national forests.

Forest Service Planning Mandates

The Forest and Rangeland Renewable Resources Planning Act (RPA), passed in 1974, established a process for comprehensively analyzing the nation's demand and supply of various resources and for setting long-range national goals to guide the annual budget process. The National Forest Management Act (NFMA), passed as an amendment to RPA in 1976, prescribed land management practices and established a process for the preparation of ten-year management plans for each of the national forests. Under RPA, an Assessment of resource supply and demand is prepared by the Forest Service at ten year intervals. Every five years a Program is prepared responding to the needs identified in the Assessment and establishing both budget targets and output goals for each resource program on the national forests for the ensuing five years. A presidential Statement of Policy accompanies each new Program as it is submitted to Congress, which can accept, reject or modify the Program and policy statement. Annual budget proposals to Congress must indicate where proposed funding levels differ from the budget goals in the Program and justify those differences. The Forest Service must prepare an annual report to Congress describing its program accomplishments and the extent to which the RPA Program goals are being met.

In response to the public controversies at the time of its passage, NFMA prescribed certain national forest land management practices, such as maximum clearcut sizes and standards for determining the physical, biological and economic suitability of lands for commercial timber production. These standards were incorporated into guidelines for the preparation of detailed land management plans to be prepared by the Forest Service for each national forest. The process is similar to the standard "rational-comprehensive" planning model. Issues and concerns are identified, relevant information is collected and analyzed, and a range of alternatives is formulated and evaluated according to various socioeconomic and environmental criteria. A decision is made selecting the alternative that appears to best address the issues within the bounds of the planning criteria. The process is further refined by guidelines in the National Environmental Policy Act (NEPA), and actually goes beyond NEPA requirements in providing for public participation at nearly every step. An administrative appeals process is established to give citizens who take issue with the decision an avenue of recourse short of formal litigation. The forest plans are revised every ten to fifteen years or whenever it is determined that conditions or issues have changed significantly from when the current plan was formulated. The plans then serve as the basis for development of the agency's annual budget proposal.[5] Within overall budget constraints specified by the Washington office, the field units develop budget alternatives based on the forest plans, and these are the basis for the overall agency budget request that is submitted to the Secretary of Agriculture.

As a part of the overall RPA process, NFMA forest plans are designed to address the national goals articulated in the RPA Program while also being responsive to local needs. The data collection and analysis at the forest level establishes the physical and biological capabilities of the local resource, which is aggregated to serve as the information base for the national RPA Assessment. Management plans for each region, not called for in NFMA, were added by federal regulation in an attempt to resolve the "top-down or bottom-up" dilemma that arises in many strategic planning systems.[6] Plans developed at each regional office are meant to serve as the negotiating point between resource capabilities identified by the forests and resource demands identified nationally through RPA planning. The regional plans are also designed to implement NFMA-mandated standards and guidelines with better uniformity than if it were done on each national forest but with greater sensitivity to the wide range of conditions within the NFS than if it were done nationally.

Program Development and Budgeting

The internal Forest Service process of program development and budgeting is the process by which the goals and objectives identified in RPA Programs and the national forest plans are operationalized. The current policy direction given to Forest Service field offices is to

> Prepare program budget proposals consistent with long range direction provided by the Resources Planning Act (RPA) Program, regional guides, forest land and resource management plans, state forest resources programs and research program plans. Do not deviate from RPA goals and objectives except where justified by new legislation, changes in congressional or executive direction, or new demands and problems not anticipated in the RPA Program.[7]

Among the stated objectives of this process are (1) "to schedule annual priorities to implement long range plans," (2) "to ensure realistic program and budget information within a planning framework," and (3) "to accomplish agreed upon goals and objectives."[8] As the vehicle for developing annual work plans and then reporting actual resource management accomplishments, the program development and budget process brings into sharp focus the overall Forest Service management system that begins with program planning. The Forest Service's conceptualization of this system is illustrated in Figure 1.4.

The elegance of this conceptual model masks a number of unresolved difficulties in effectively linking the various planning and budgeting processes. The intersection between the top-down approach of RPA planning and the bottom-up approach of forest planning under NFMA is still incomplete and poorly defined. The regional guides have not been as useful in reconciling these two processes as had been hoped and have been deemphasized in actual practice. A more serious disjunction persists between these planning processes and the budget development process. For a variety of reasons that will be discussed below, Forest Service budgets and congressional appropriations have failed to accurately reflect the goals established through program planning. The report of National Forest System accomplishments compared to RPA goals in a recent Forest Service annual report showed that some programs had achieved as little as 21 percent of the planned outputs.[9] Much of this difference has been attributed to funding being inadequate to carry out the planning goals.[10] What may be more disturbing is that the inadequacies in funding were not uniform, as might be expected in a time when federal spending is declining generally, but represent a wholly different emphasis among the National Forest System resource programs from what was decided through RPA planning. The same Forest Service annual report referred to above indicated that the proportion of the RPA goal achieved varied across programs from the low of 21 percent to as high as 119 percent.[11] If the budget and planning processes are indeed linked, the fact is not apparent from these outcomes. The efficacy and cost effectiveness of doing such planning at all has been called into question, and more than one reviewer has declared RPA planning to be pointless and moribund.[12]

Figure 1.4
The Forest Service Management Model

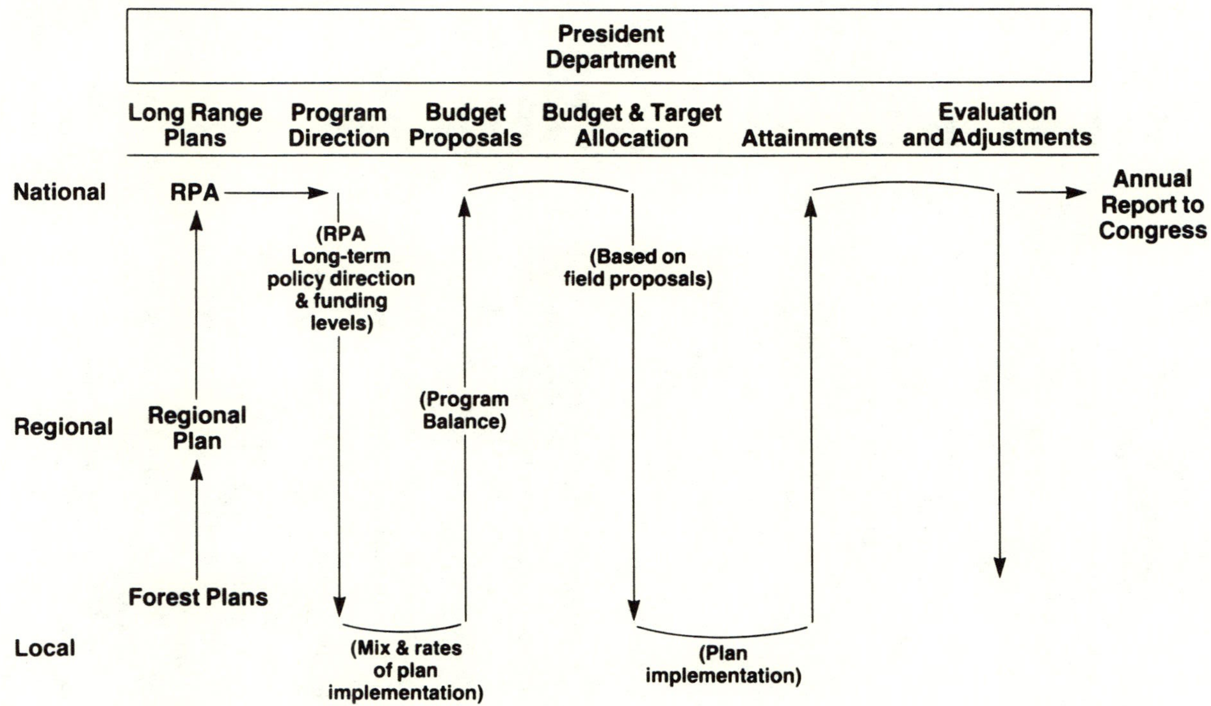

Since the National Forest Management Act became law in 1976, a tremendous amount of time and money has gone into the development of land and resource management plans for the 156 national forests (consolidated into 126 forest plans). This has involved not only the efforts of thousands of professional resource managers in the Forest Service, but also many citizens and organizations with an interest in how the national forests will be managed over the next decade. As most of the plans near completion and estimates are made of the cost of implementing them, it is clear that a significant increase in National Forest System appropriations would be required (see Table 1.1). Just as clearly, the size of the federal budget deficit will continue to require restraint, if not major cutbacks, in federal spending.[13] What will be the impacts of the federal budget process on the implementation of these plans? Can we expect the actual management of the national forests to at all reflect the programs and priorities selected during the long years of plan development, public review, appeals and litigation? Can the Forest Service's budget development process be any better integrated with the implementation of national forest planning than it was with RPA Program implementation?

Study Approach

The objective of this study was to address these questions by gaining a more thorough understanding of what actually happened with Forest Service budgeting during the first decade of the implementation of RPA, exploring the reasons why events and decisions transpired as they did, and examining the critical points where NFMA planning can be better integrated with the realities of the budget and appropriations processes than RPA has been. The first step was to examine in detail the appropriations provided over the past ten years for each resource program within the National Forest System and compare them to the funding goals depicted in the three RPA Programs prepared to date. However, appropriations do not, like Athena from the forehead of Zeus, spring forth in full adulthood and ready for battle. To understand how and why the appropriations emerged as they did, it is necessary to explore the internal Forest Service mechanisms by which planning and budgeting decisions are actually made; it is also necessary to explore the dynamics of governmental organizations external to the Forest Service that exert varying degrees of control over the agency's activities. These organizations include the Office of the Secretary of Agriculture, the Office of Management and Budget, and the House and Senate appropriations committees.

The Forest Service, though it enjoys extraordinary autonomy and independence in comparison to most federal agencies, hardly exists in a vacuum. In order to provide a more thorough understanding of the underlying forces directing the management of the

Table 1.1
Actual Budget Trends and Plan Implementation Requirements - National Forest System

RESOURCE PROGRAM	FY 1983	FY 1984	FY 1985	FY 1986	FY 1987	FY 1988	FY 1989	FOREST PLANS FY 1989	PERCENT OF FY 89 FOREST PLANS
	[THOUSANDS OF NOMINAL DOLLARS]								
TIMBER	813.6	772.7	798.0	733.4	740.9	813.6	847.0	971.1	87%
RECREATION	121.3	122.7	129.9	126.3	149.4	180.0	171.1	288.4	59%
FISH & WILDLIFE	61.1	58.1	60.9	63.3	73.7	92.3	86.2	131.0	66%
RANGE	32.2	32.2	32.0	30.7	31.7	33.3	30.5	48.2	63%
SOIL & WATER	28.1	29.7	31.0	30.5	34.2	37.1	34.3	56.0	61%
MINERALS	22.1	25.4	26.5	27.2	26.9	28.1	30.0	41.1	73%
PROTECTION	150.1	154.9	157.3	151.7	158.6	164.7	154.0	200.1	77%
FACILITIES	113.5	115.3	115.6	109.4	99.9	117.9	103.6	179.5	58%
LANDS	100.5	86.3	93.4	78.7	100.3	99.2	56.9	90.1	63%
TOTAL	1,442.5	1,397.3	1,444.4	1,351.2	1,415.7	1,566.1	1,513.6	2,005.5	75%

Source: USDA Forest Service, Land Management Planning

national forests, the analysis explicitly recognizes the interplay of these external organizations with the Forest Service and between themselves with regard to Forest Service activities. The outcome of the budget and appropriations process for the Forest Service for a single fiscal year, fiscal year (FY) 1986, will be explored from the perspectives of three alternative conceptual models developed by Graham Allison.[14] The outcome is explored first as the product of a rational policy process with the various decision makers performing essentially as a single rational actor; second, the outcome is viewed as the output of several distinct organizations within the federal government, each acting according to its own value system and standard operating procedures; third, the outcome is seen as the product of the strategizing and bargaining that characterize bureaucratic politics within and between executive and legislative branch organizations in the federal government. The explanatory power of each different model is discussed and evaluated in the specific context of the people and events associated with Forest Service appropriations for fiscal year 1986.

The decisions critical to determining how the national forests are actually managed do not end with the signing of an appropriations bill, but continue throughout the execution of the budget approved by Congress. Since the focus of this study is the implementation of program plans prepared by field administrators for 156 national forests scattered from Puerto Rico to Alaska, the important questions become: In what ways did actual appropriations differ from the budget proposals prepared at the field level on the basis of the national forest land and resource management plans? To what extent did these differences limit the ability of local level managers to carry out the activities anticipated in the budget proposals and forest plans? Thus, we will examine the budget allocation and execution processes used by the Forest Service, including its reporting back to Congress of detailed accounting information regarding both the agency's expenditures and its accomplishment of the resource management objectives assigned by Congress in the context of the appropriations bill. This will be accomplished through case studies of five different national forests, each in a different region of the country and each facing a different set of resource management issues and opportunities. The budget proposals and planned management activities for each case study national forest will be tracked (1) throughout the development of the overall Forest Service budget request, (2) through the allocation of funding and resource management targets from Congress back to each national forest, and (3) to the end-of-year reporting of expenditures and actual target accomplishments. The comparison of funding and resource management activities at these three stages should give some indication of where changes between planned and actual budgets became limiting and where they did not. Because a simple comparison of such numbers often does not tell the complete story, follow-up interviews with resource managers on each case study national forest helped to guide in the interpretation and evaluation of the comparisons.

Both budgeting and program planning are necessary processes for any large and complex organization, especially one as decentralized as the Forest Service. The ideal is for these two processes to be well integrated, and to be supportive of and complementary to one another. In many organizations, however, they seem to operate at cross purposes, perhaps because planning--especially that done at the local level--is largely done by program specialists concerned with particular activities the organization undertakes and budgeting is done largely by process specialists concerned with accountability and the smooth functioning of the organization as a whole. While some flexibility for adjusting appropriations to unforeseen management needs has been built into the congressional appropriations control system, it is unclear whether or not the flexibility provided is either useful or sufficient for that purpose. In the final section, opportunities are explored for modifying the current processes, both program planning for budget development and appropriations, to achieve a more effective integration of the two, thereby improving the prospects for the Forest Service's actually accomplishing the goals and objectives set out in the first generation of forest plans developed to implement the purposes of the National Forest Management Act.

NOTES

1. Humphrey, Hubert H., Floor statement to accompany introduction of S. 2296, "The Forest and Rangelands Renewable Resources Planning Act," cited in: Wolf, Robert E., "Architects and Architecture," *The RPA Process: Moving Along the Learning Curve*, Gerald E. Stairs and Thomas E. Hamilton, eds. (Durham, N.C.: Duke University Press, 1983).

2. U.S. Department of Agriculture, *Budget Estimates for the U.S. Department of Agriculture for the Fiscal Year Ending September 30, 1986* (Washington, DC: Government Printing Office, 1985).

3. Ibid.

4. This describes the national forests of the continental United States. The Tongass National Forest in Alaska comprises 16.8 million acres, nearly five times the size of the state of Connecticut.

5. 36 CFR (Code of Federal Regulations) 219.10(e).

6. 36 CFR 219.4(b)(2) and 36 CFR 219.9. For further discussion of the problem of top-down or bottom-up approaches to strategic planning, see generally: Naylor, T. H., *Strategic Planning Management*

(Oxford, OH: Planning Executives Institute, 1980).

7. USDA Forest Service, *Forest Service Manual*, Title 1930.2, as amended September, 1985 (Washington, D.C.: USDA Forest Service, 1985).

8. Ibid.

9. USDA Forest Service, *Report of the Forest Service, Fiscal Year 1985* (Washington, D.C.: USDA Forest Service, 1986), p. 80.

10. Sample, V. Alaric, "The Resources Planning Act (RPA) Process and Its Implementation to Date," in: Binkley, C. S., Brewer, G. D. and Sample, V. A. (eds.), *Redirecting the RPA: Proceedings of the 1987 Airlie House Symposium on the Resources Planning Act,* Bulletin No. 95 (New Haven, CT: Yale School of Forestry & Environmental Studies, 1988).

11. USDA Forest Service, *Report of the Forest Service, Fiscal Year 1985*, p. 80.

12. See: Behan, Richard W., "RPA/NFMA: Time to Punt," *Journal of Forestry*, Vol. 79 (1981); Leman, Christopher K., "Planning Against Analysis: Forest Service Implementation of the Resources Planning Act," in: Binkley, C. S., Brewer, G. D. and Sample, V. A. (eds.), *Redirecting the RPA: Proceedings of the 1987 Airlie House Symposium on the Resources Planning Act,* Bulletin No. 95 (New Haven, CT: Yale School of Forestry & Environmental Studies, 1988); Sample, V. Alaric, "The Rise and Fall of the Resources Planning Act: 1974-1985," Lecture in Natural Resources Policy (unpublished paper), Yale University, School of Organization and Management, New Haven, CT, 1986.

13. LeMaster, Dennis C., "Federal Budgeting, Gramm-Rudman-Hollings, and Natural Resource Program Funding," in: *Proceedings of the Society of American Foresters Annual Meeting, 1986* (Bethesda, MD: Society of American Foresters, 1987).

14. Allison, Graham T., "Conceptual Modes and the Cuban Missile Crisis," *American Political Science Review*, Vol. 53 (1969) pp. 689-718.

FOREST SERVICE BUDGET DEVELOPMENT

2

Forest Service Budgets and Appropriations: An Historical Analysis

The analysis of Forest Service budget development was conducted in two phases. First, a 25-year time series was developed showing the funding proposed for each National Forest System resource program at each stage of the appropriation process. For fiscal years 1977 to 1986, the prior year's appropriations, or "base," is shown as a point of reference, followed by the Forest Service agency request, the proposal by the Office of the Secretary of Agriculture, the President's Budget (OMB proposal), action on the budget by the House of Representatives, action by the Senate and the final appropriation agreed upon in House and Senate conference and signed by the president. Data for the Presidents' Budgets and congressional budget actions, by NFS resource program, were compared for fiscal years 1960 through 1986 (budget requests from the Forest Service and from the Department of Agriculture are no longer available for fiscal years prior to 1977). The purpose of this analysis was to explore modifications made by each participant in the budget process and to identify any regular or systematic actions on the part of any of the various participants in the process. The time series was also useful in examining the correlation between the proposals of each participant with actual appropriations. Such information also should reveal the relative effectiveness of each participant, particularly those in the executive branch, in influencing the final outcome of the budget process. For fiscal years 1977 to 1986, the annual budget targets from the 1975 and 1980 RPA Programs were also included, both as a point of reference when examining the budget proposals of each participant in the process and to determine the degree of congruency between the RPA budget targets for NFS programs and actual appropriations.

The second phase of the research involved a series of case studies conducted on five national forests around the country to examine the details of budget development for a single fiscal year, 1986. The purpose of the case studies was two-fold, encompassing not only budget development but budget execution as well. First, the budget proposal was tracked through each level of decision making--from the budget proposal developed by field officials at each national forest up through the final appropriation by Congress --in order to explore the priorities, policy thrusts and ultimate influence of each participant. The appropriation was then tracked back down through the administrative levels of the Forest Service itself to examine the influence of internal processes on the allocation of funds to each case study forest. The purpose of this analysis is to explore the nature, degree, and source of any modifications made to the funding and resource output targets in the original field budget proposals. Second, following the execution of the budget through not only the fund allocation to the field, but also to the actual expenditures and accomplishments reported at the end of the fiscal year, made it possible to assess the ability of field officials to carry out program plans within the constraints of a budget that often had been modified in terms of both overall funding and the distribution of funding among resource programs. A series of follow-up interviews with the field officials responsible for implementing the fiscal year 1986 program on each of the case study forests permitted a further exploration of the degree of flexibility available to field officials to carry out necessary management activities within the legal and administrative constraints of their budget allocations.

Forest Service administrative records and congressional process documentation were used to reconstruct actions on the Forest Service budget for fiscal years 1960 through 1986. Information provided by the Forest Service office of Program Development and Budget (PD&B) was checked against an historical data base maintained by OMB and against records of the appropriations committees in both the House and Senate. Information concerning budget proposals from the Forest Service and the Department of Agriculture is difficult to obtain, and few official records of the Forest Service/Agriculture/OMB negotiations leading up to the OMB budget proposal for the Forest Service in the President's Budget have been retained. This is due in part to long-standing direction from OMB to federal agencies requiring that "executive branch communications that have led to the budget will not be disclosed either by the agencies or by those who have prepared the budget."[1] According to this direction, which has the force of a presidential executive order, agencies are forbidden to disclose agency or departmental budget proposals leading up to the President's Budget to Congress or the public--until the end of the fiscal year to which that budget applies. After the end of the fiscal year, information from the preliminary budget proposals may be released only in response to a direct request under the Freedom of Information Act (FOIA) (5 U.S.C. 552 (b)(5)). Even so, significant portions of this body of information have been declared exempt from FOIA requests and thus can be permanently withheld from release to the public.[2] For fiscal years prior to 1977, even the Forest Service and Department of Agriculture do not maintain an official record of the requests that led up to the proposals in the President's Budget. Beginning with fiscal year 1977, this information has been maintained for internal use by the Forest Service but has not been made generally available to the public.

This study will thus focus primarily on Forest Service planning and budgeting for fiscal years 1977 to 1986. An understanding of the historical context of this era will be important throughout the interpretation of the budget information. This was a time of great change for the Forest Service and for federal programs overall. National concerns over high inflation during the late 1970s forced a re-evaluation of federal monetary and fiscal policies. Paul Volcker's replacement of Arthur Burns as chairman of the Federal Reserve Bank ushered in an era of tighter money supplies and rising interest rates, successfully bringing down the inflation rate. These policies had a marked effect on interest rate-sensitive industries, however, chief among them the housing and lumber industries. As interest rates rose, the demand for softwood lumber and plywood plummeted from near peak activity in the late 1970s to its lowest level in decades in 1982 and 1983, recovering only gradually in the years that followed. With the national forests typically providing more than a quarter of all the softwood timber harvested in the United States, these trends had important implications for the Forest Service.

Both Congress and the executive branch were faced with changing priorities as a result of these economic shifts. The tide that swept the Reagan administration into power in 1981 also gave the Republican party a majority of seats in the U.S. Senate, leading to changes in leadership, committee chairmanships, and agendas. It also complicated the relationship of Senate leaders with the House--whose leaders remained Democratic--especially in terms of appropriations priorities and strategies. The social and political context cannot be held constant over time, and the time-series data presented here must be explored against this shifting backdrop.

For analysis purposes, sixty or so budget line items in the annual proposals were aggregated by NFS resource program: timber, minerals, recreation/wilderness, fish and wildlife, range, soil and water, protection, and facilities. This aggregation was performed consistent with a "crosswalk" developed by the Forest Service to translate the multitude of budget line items into the resource programs that are the basis for analysis and display in the RPA documents.[3] Such an aggregation largely eliminates the variation caused by the numerous and often poorly documented changes that have been made in the nomenclature for budget line items over the 27 years studied. It also captures reasonably well the contribution made by administrative and capital investment costs to the overall costs of individual resource programs when, under the sort of integrated resource management that characterizes the national forests, these overhead and investment expenditures serve more than one program. Beginning in fiscal year 1982, Congress required the Forest Service to show its costs for general administration (line officer and administrative staff expenses, office rent and operation) as a separate budget line item rather than apportioning these costs among the various resource programs. To make pre-1982 program costs consistent with those in subsequent years, general administration costs were factored out of the pre-1982 figures with guidance from officials at OMB. Finally, all figures were converted to constant 1982 dollars using the Gross National Product (GNP) implicit price deflators for 1960 through 1986.[4]

Data for fiscal years 1977-1986 was expanded to include the budget goals appearing in the 1975, 1980 and 1985 RPA Programs. As noted above, this permits the comparison of various budget proposals and actual appropriations in each year to the RPA targets. Thus, one can assess the degree of influence that comprehensive program planning had on both the level and distribution of Forest Service funding, and whether there is any pattern in the deviation from these targets among individual resource programs.

Analysis of Budget Proposals for National Forest System Resource Programs

The time-series data are illustrated in the series of graphs below. The cluster of bars for each fiscal year, moving from left to right, indicates the sequential decisions made by the various participants in the budget process as each responded to the previous participants' proposals and/or the prior year's actual appropriation. The RPA budget targets for each resource program, as shown in the RPA Programs for 1975, 1980 and 1985, are included for reference. A somewhat more quantitative analysis of the data is provided through statistical correlations between the proposals of the various participants in the budget process and final appropriations.

National Forest System

Beginning with FY 1977 (see Figure 2.1) the Forest Service seems to have pursued an aggressive strategy for the National Forest System, requesting considerable increases over the prior year base. Up through FY 1982, this strategy appears to have been successful in significantly boosting the agency's appropriations in real terms, though not by as much as the Forest Service requested. For FY 1977-1986, the Forest Service request was an average of 23 percent over its previous year's appropriation for National Forest System programs overall; appropriations actually increased by an average of 6.5 percent annually over the period (see Table 2.1). Increases are heavily concentrated toward the late 1970s, however. In FY 1977 the Forest Service showed remarkable confidence in its relationship with Congress by requesting a 50 percent increase over the previous year's funding. The effort was as successful as it was optimistic and the agency actually received a 41 percent increase in appropriations for that year alone.[5] The Forest Service prepared its FY 1978

Figure 2.1
Forest Service Budgets, FY 1977-1986
National Forest System - total

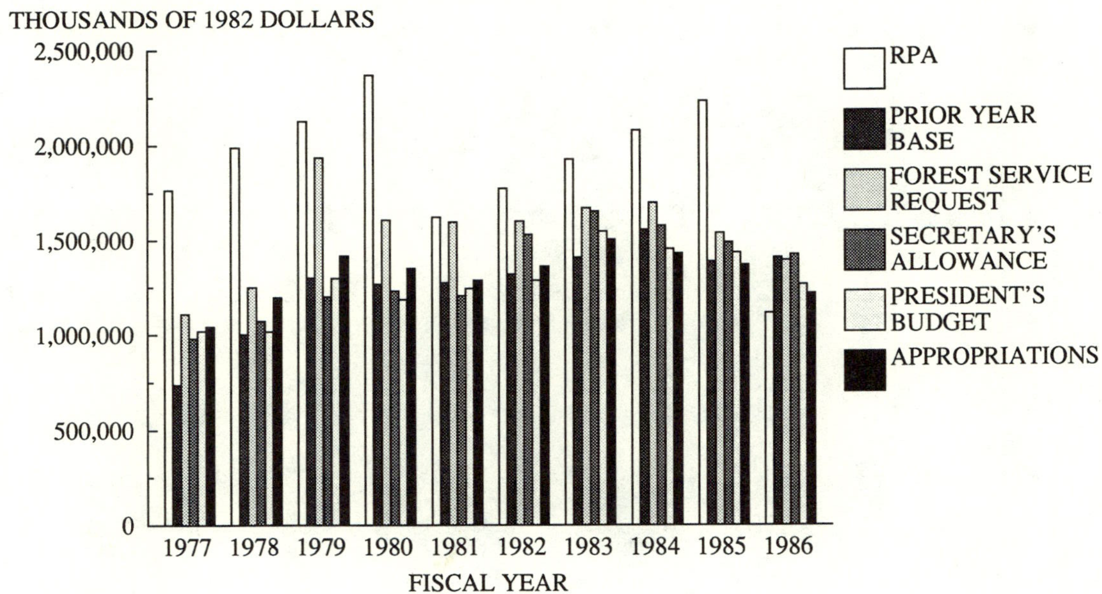

Table 2.1
Average Percent Change of Budget Proposals from the Prior Year Base, by Participant and Resource Program, FY 1977-1986

PROGRAM	TIME PERIOD	RPA	USFS	USDA	OMB	HOUSE	SENATE	APPROPRIATIONS
				[PERCENT]				
TIMBER	1977-1986	19.8	16.7	8.1	3.0	2.7	5.7	5.9
	1977-1981	37.8	22.5	5.2	5.8	15.6	14.7	15.3
	1982-1986	1.7	11.0	11.1	0.3	-10.2	-3.3	-3.5
RECREATION	1977-1986	102.4	31.2	-1.3	-6.9	5.5	-0.9	5.3
	1977-1981	132.8	55.6	1.3	-0.4	15.3	9.3	13.1
	1982-1986	72.0	6.8	-3.9	-13.3	-4.4	-11.2	-2.5
WILDLIFE & FISH	1977-1986	86.1	53.7	10.8	2.6	18.6	15.8	15.3
	1977-1981	80.7	87.5	14.3	7.6	37.4	28.8	30.4
	1982-1986	91.5	19.9	7.4	-2.3	-0.2	2.7	0.1
RANGE	1977-1986	61.3	24.3	1.4	-4.4	3.9	5.9	3.7
	1977-1981	73.9	38.7	2.3	-2.7	15.1	10.2	11.0
	1982-1986	48.7	10.0	0.4	-6.0	-7.4	1.6	-3.6
SOIL & WATER	1977-1986	38.5	25.6	-4.8	-6.9	5.4	-0.2	3.8
	1977-1981	1.6	35.7	-4.5	-5.4	10.8	4.2	8.2
	1982-1986	75.4	15.4	-5.0	-8.4	-0.0	-4.7	-0.6
MINERALS	1977-1986	*	42.0	26.3	19.2	15.0	18.9	15.0
	1977-1981	*	55.0	24.4	15.6	18.0	16.6	17.9
	1982-1986	47.3	29.1	28.2	22.8	12.1	21.2	12.2
PROTECTION	1977-1986	*	46.8	4.0	17.1	8.3	17.3	17.9
	1977-1981	*	85.8	1.8	35.1	15.6	32.5	34.7
	1982-1986	28.0	7.7	6.2	-0.9	1.0	2.1	1.0
LANDS	1977-1986	*	50.4	10.7	6.7	18.0	11.6	13.5
	1977-1981	*	87.1	17.3	10.9	32.6	20.4	24.2
	1982-1986	312.0	13.7	4.1	2.6	3.4	2.7	2.8
FACILITIES	1977-1986	*	36.7	8.1	0.9	4.9	3.6	4.6
	1977-1981	*	51.4	6.0	3.1	13.9	9.1	12.3
	1982-1986	63.1	22.1	10.2	-1.4	-4.1	-1.9	-3.1
NFS-TOTAL	1977-1986	*	23.5	6.7	2.4	4.3	5.6	6.5
	1977-1981	*	35.2	4.9	6.1	15.6	14.0	15.5
	1982-1986	29.0	11.8	8.6	-1.2	-7.1	-2.8	-2.6

* Not available. The 1980 RPA Program included budget targets for programs not included in the 1975 Program. The totals for the 1975 Program are thus not directly comparable.

budget request before its phenomenal success with the FY 1977 request could be known and proposed only a relatively modest additional increase. The FY 1979 budget was based on FY 1977 actual funding, however, and apparently on a realization that bold initiatives could bring substantial reward. The Forest Service proposed an additional 49 percent increase in NFS funding (see Table 2.2). However, whereas the FY 1977 initiative had enjoyed the support of the secretary of agriculture, who proposed a 33 percent increase, and OMB, which proposed a 38 percent increase, this support was clearly lacking for FY 1979 when both the secretary and OMB called for a budget decrease in real terms for the National Forest System programs. Congressional support ultimately provided a more modest, but nonetheless substantial, nine percent increase in real terms. Significant real increases continued until the budget for FY 1982, the first budget prepared fully by the Reagan administration and reviewed by a Republican majority in the Senate. A long downward trend in both requests and funding culminated in FY 1986, when the agency proposed a one percent real decrease in NFS funding and ultimately emerged with a reduction of more than thirteen percent.

Timber

The timber program is focused primarily on the preparation, sale and harvest of between ten and twelve billion board feet of timber annually from the national forests. It also provides for the salvaging of dead and damaged timber following fires and insect or disease infestations, usually another billion board feet or more annually. In addition, timber staff officers oversee the sale of "non-convertible products" such as fuelwood and Christmas trees. The reforestation and timber stand improvement (TSI) portion of the timber program includes the operation of nurseries to provide seedlings for reforesting harvested or burned over areas, selective breeding programs for the genetic improvement of nursery stock, site preparation in advance of direct seeding or natural regeneration of harvested areas, and thinning and other treatments to improve growth rates in immature stands. To allow for the removal of logs and other wood products, the timber program also includes the construction of many miles of secondary roads and "skid roads," which provide direct truck access to timber sales.

The timber program constitutes nearly two-thirds of National Forest System funding, and trends in the timber budget are thus similar to those for the NFS overall. Direct appropriated funding for reforestation/TSI and other post-harvest activities is augmented significantly by deposits made in advance by timber purchasers. The Knutson-Vandenberg (K-V) Act of 1930 (16 U.S.C. 576-576b) and the Brush Disposal Act of 1916 (16 U.S.C. 490) each permit the Forest Service to collect deposits for use in rehabilitating timber sale areas after harvesting. Throughout this analysis, figures indicating

Table 2.2
Percent Change in Forest Service Budget Proposal From the Prior Year Base, by Participant, FY 1977-1986, National Forest System

FISCAL YEAR	PRIOR YEAR BASE	RPA	USFS	USDA	OMB	HOUSE	SENATE	APPROPRI-ATIONS
	[1982$]			[PERCENT]				
1977	738,365	*	50.5	32.9	38.2	42.4	44.6	41.5
1978	1,001,919	*	24.9	7.3	1.5	22.3	17.3	19.5
1979	1,301,191	*	48.7	-7.7	-0.2	4.9	4.6	9.0
1980	1,270,036	*	26.6	-3.0	-6.7	6.1	4.6	6.4
1981	1,274,573	27.2	25.2	-5.3	-2.4	2.2	-1.0	1.1
1982	1,319,461	34.5	21.4	15.9	-2.4	-0.4	3.8	3.2
1983	1,409,133	36.8	18.5	17.4	9.8	3.7	6.1	6.7
1984	1,555,783	33.7	9.1	1.4	-6.6	-9.5	-4.2	-8.0
1985	1,388,267	60.9	10.9	7.2	3.3	-4.0	3.9	-1.3
1986	1,408,485	-20.8	-1.1	1.2	-10.1	-25.1	-23.4	-13.4

* Not available. The 1980 RPA Program included budget targets for programs not included in the 1975 Program. The totals for the 1975 Program are thus not directly comparable.

funding levels for the timber program include the authorized expenditures from these deposits.

Because the timber program budget is much larger than those of other resource programs it fluctuates much more in absolute terms than other resource programs. The most vivid example of this is the study year of FY 1986, when the budget for the timber program varied from the secretary's request of a 4 percent increase over the prior year base to the House's proposal of a 38 percent *reduction* from the base (see Figure 2.2). This amounted to a difference of nearly $400 million.

When compared to the other resource programs, appropriations for the timber program followed fairly close to the budget goals set in the RPA Programs. A fairly aggressive budget strategy by the Forest Service during the late 1970s received decidedly lukewarm support from the secretary of agriculture and from OMB. For FY 1978-1981, the Forest Service proposed an average 13 percent increase over the prior year base while the secretary proposed an average 4 percent *reduction* from the base and OMB proposed an average 6 percent reduction. The timber program received greater emphasis from Congress during this period, however, and actually received an average 5 percent annual increase in real terms in the final appropriations. With the change in presidential administrations in time for FY 1982 budget development, this trend was largely reversed. For the fiscal years 1982-1986, the Forest Service request for the timber program averaged an 11 percent increase over the base (see Table 2.1), matching the average 11 percent increase in the secretary's request. The change at OMB was not nearly so pronounced, its average budget proposal for the period moving up to only .3 percent above the base. Despite a new conservative majority in the Senate, the Congress overall seemed to reverse its previous role of adding back funding that the secretary and OMB had proposed to cut from the Forest Service request. For FY 1982-1986, congressional appropriations for the timber program averaged a 3 percent reduction from the base (culminating in FY 1986 with a 21 percent reduction from the previous year's funding), well below the Forest Service requests and below even OMB's proposals.

This declining trend reflects not only the general cutbacks in funding for federal programs in an effort to control the growing federal budget deficit, but reductions in national forest timber program funding in particular in response to the 1982-1983 economic recession. During this period, many timber sales purchased during the late 1970s became economically inoperable and were repurchased by the Forest Service to avoid a mass default by timber purchasers. When these same timber sales were reoffered in FY 1985 and 1986, costly investments such as road access had already been made, thus the per-unit cost to the Forest Service of these reoffered sales was lower than that of new timber sales. Figure 2.3 shows the actual level of national forest timber offerings, sales and harvests relative

Forest Service Budgets and RPA Program Targets
Fiscal Years 1977 - 1986

Figure 2.2
Timber Program

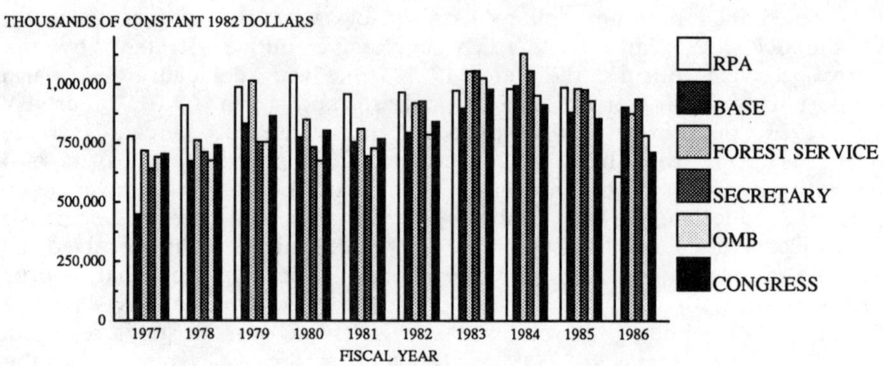

Figure 2.3
Comparison of RPA Projected Demand for National Forest Timber with Actual Use, FY 1977-1986

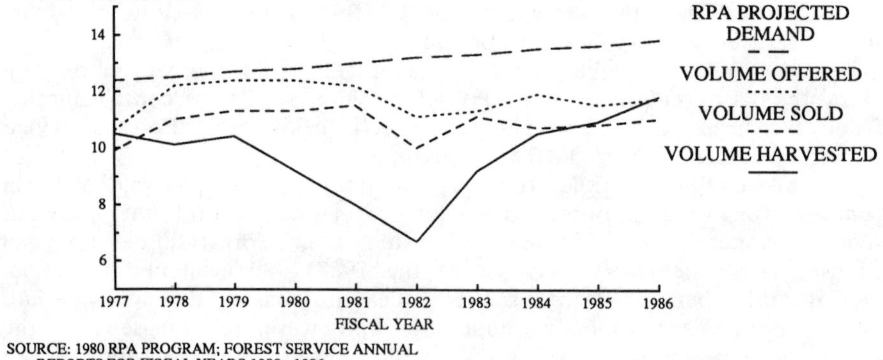

SOURCE: 1980 RPA PROGRAM; FOREST SERVICE ANNUAL REPORTS FOR FISCAL YEARS 1980 - 1986

to levels projected in the RPA Assessments. The volume of timber offered did fall off somewhat from the peak years of FY 1979 and 1980, but the reduction in volume offered following FY 1984 is insignificant when compared to the reduction in timber funding. The program level thus remained largely unchanged despite the apparent decline in appropriations.

Recreation and Wilderness

More recreation use occurs on the national forests than on any other federal, state or private lands, generally about twice that in the national parks. The recreation program includes the construction and maintenance of developed recreation sites such as campgrounds, picnic areas and boat launches as well as facilities for dispersed recreation use, such as trails and backcountry campsites. The recreation program also includes the responsibility for the management and protection of the 62 million acres of congressionally designated wilderness on National Forest System lands.

Congressional appropriations for recreation and wilderness were far below the funding levels recommended in the RPA Programs. Forest Service budget requests for its recreation program during the first two years of the 1975 RPA Program (FY 1977 and 1978) called for a relatively modest increase over the prior year base and are well below the budget targets in the RPA Program (see Figure 2.4). Again, because the Forest Service in any given year is engaged in developing its budget request two years in advance, agency budget strategists receive important cues from the current year's actual appropriations that often show up in subsequent agency requests, but with a two-year lag. For example, the recreation program shared in the large overall increase in FY 1977 appropriations, thus the FY 1979 agency request for this program, too, represented a bold increase (63 %) over the FY 1978 base. Significant real-term increases in appropriations for recreation for fiscal years 1978 and 1979 similarly encouraged a 69 percent and 73 percent increase in the agency requests for FY 1980 and 1981, respectively.

This aggressive strategy on the part of the Forest Service was pursued without the benefit of any apparent support from the secretary of agriculture or OMB, however. For FY 1977-1981, the agency request for the recreation program averaged 56 percent above the prior year base; the secretary's request for this program over the same period averaged only 1.3 percent above the base, and OMB proposed an average .4 percent *reduction* from the base. With the change in administrations, even the Forest Service strategists appear to have been reined in; agency requests for the recreation program for fiscal years 1982-1986 averaged only 7 percent above the base, having fallen sharply between the FY 1981 request and the FY 1982 request, then again between the FY 1983 and FY 1984. The secretary's requests for this program averaged 4 percent below the base for this

Forest Service Budgets and RPA Program Targets
Fiscal Years 1977 - 1986

Figure 2.4
Recreation Program

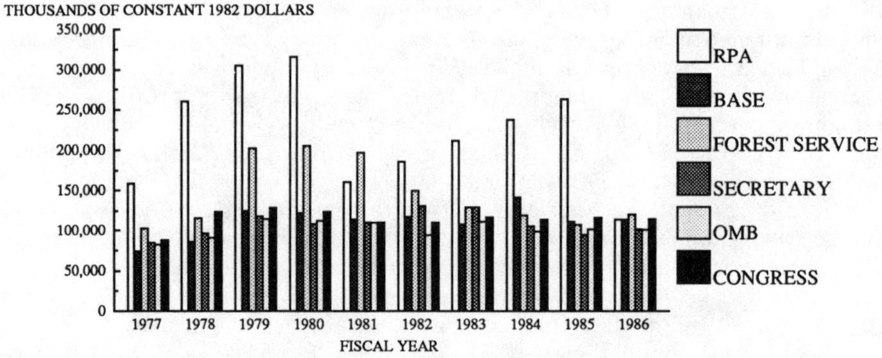

Figure 2.5
Comparison of RPA Projected Demand for National Forest Recreation with Actual Use, FY 1977-1986

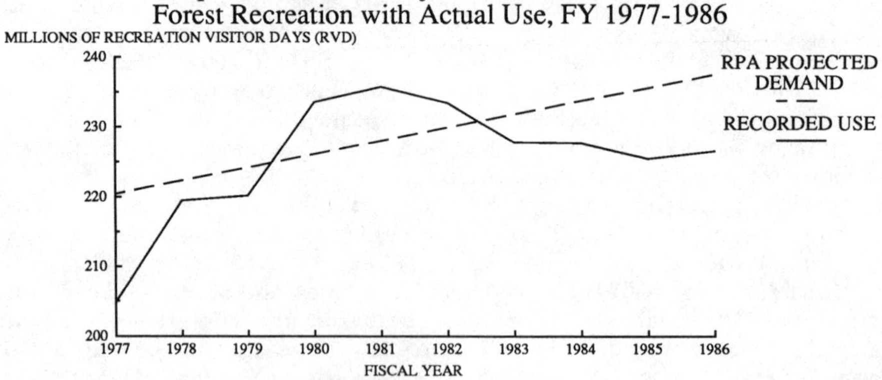

SOURCE: 1980 RPA PROGRAM; FOREST SERVICE ANNUAL REPORTS FOR FY 1980-1986.

period, and the OMB proposals averaged 13 percent below the base. Very sharp drops in support for this program for FY 1984 by the secretary, OMB and even the Congress may be related to declines in the reported recreation use levels on the national forests following a peak in FY 1981 (see Figure 2.5). Evidence of this declining trend, which is sharply at odds with the demand trend projected by the Forest Service in the 1979 RPA Assessment, appeared in the agency's FY 1982 report to Congress. This information would have become available in the fall of 1982, just as USDA and OMB budget analysts were reviewing the Forest Service's request for FY 1984, and would have been officially transmitted to Congress just prior to appropriations committee hearings on the FY 1984 President's Budget for the Forest Service.

Fish and Wildlife, Range

The Forest Service is responsible for managing wildlife and fish habitat on the national forests, while the state governments are responsible for managing the animal populations on these lands. This includes not only the maintenance of adequate forage and cover for a variety of game and non-game birds, but also spawning habitat for anadromous fish in the headwaters on some of the most productive fisheries in the nation. While most of the wildlife program is focused on the enhancement of existing habitat, especially for threatened and endangered species, a major portion of the program is focused on the mitigation of timber harvesting activities on wildlife habitat and streams. Timber receipts withheld from the Treasury under provisions of the Knutson-Vandenberg Act comprise a large share of the funding for mitigation and coordination activities. The range program is focused on maintaining and improving the productivity of national forest lands for livestock grazing. Grazing permits are issued at a nominal fee (based on the number of animals grazed) and the level of use is varied depending on land productivity and conditions.

The pattern of Forest Service budget requests for both the fish and wildlife program and the range program, and the response of each subsequent participant in the budget process, was very similar to that for the recreation program (see Figures 2.6 and 2.7). A very aggressive Forest Service budget strategy during the late 1970s was successful in obtaining substantial increases in congressional appropriations, this despite little apparent support from either the secretary of agriculture or OMB. During this period, the budget for the fish and wildlife program grew at a faster rate than any time in its history and at a faster rate than any other National Forest System resource program, more than doubling from FY 1977 to FY 1981. This aggressive Forest Service strategy was sharply moderated beginning in FY 1982, and even modest attempts by the agency to increase funding for these programs through FY 1986 met with little enthusiasm

Forest Service Budgets and RPA Program Targets
Fiscal Years 1977 - 1986

Figure 2.6
Fish and Wildlife Program

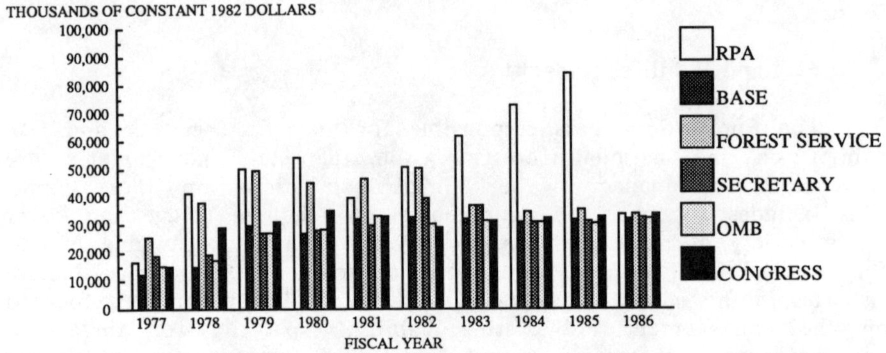

Figure 2.7
Range Program

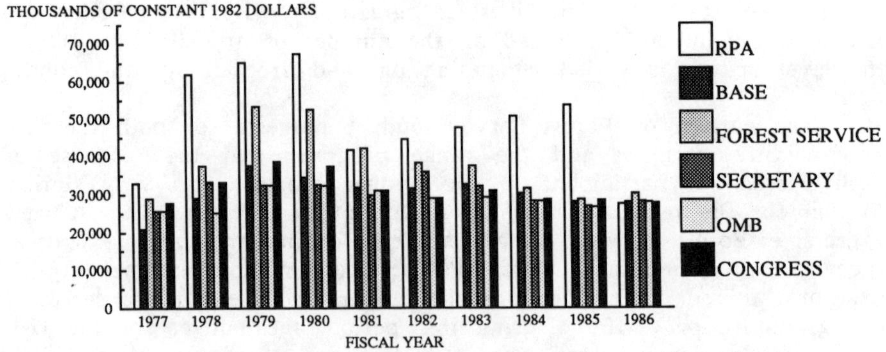

elsewhere in the executive branch. Congress, under increasing pressure to control federal spending, seems to have mechanically split the differences between the Forest Service and OMB for the fish and wildlife program. By FY 1986, all participants seem to have been in virtual lock-step with one another, with few modifications being proposed along the way, not only for these programs but for every National Forest System resource program except timber.

Soil and Water

Nearly half of the expenditures for the soil and water program are for mitigating the effects of other resource management activities, particularly timber harvesting, on soil erosion, non-point source water pollution and riparian values. Deposits under the Knutson-Vandenberg Act form a major share of this funding. The program also provides for soil inventories to provide information on soil productivity, erosion and stability problems, and for soil stabilization and stream protection projects to improve watershed conditions. The program also includes emergency rehabilitation, through grass seeding and other erosion control measures following floods and fires (often through additional funding measures such as the Agricultural Credit Act of 1978). Reclamation of abandoned surface mines is often undertaken with additional funding through the Surface Mining and Control Act.

The soil and water program has its own unique pattern for FY 1977-1981 (see Figure 2.8). The Congress was very supportive of the soil and water program, in some cases significantly exceeding the Forest Service budget request, as it charged the Forest Service with a multitude of new monitoring, analysis and reporting responsibilities under the National Environmental Policy Act, the Clean Water Act and the Federal Water Pollution Control Act. However, whereas the funding for many of the other nontimber resource programs was ratcheted upwards during the late 1970s and then held relatively constant near that plateau through the mid-1980s, funding for soil and water programs has fallen significantly from FY 1980 levels. The Forest Service requests relative to the prior year base have declined steadily, as have those from the secretary and OMB. Again, Congress appears to have mechanically split the differences between the agency and OMB in its final appropriations from FY 1982 onwards.

Minerals

Large quantities of "leasable" minerals (e.g., coal, oil, gas, geothermal steam) and "locatable" minerals (e.g., uranium, chromium, nickel, gold, silver, copper and molybdenum) are found beneath national forest lands. The Forest Service minerals program ensures that the mineral resources are developed in a manner compatible with the management of the surface resources like timber, wildlife and water.

Forest Service Budgets and RPA Program Targets
Fiscal Years 1977 - 1986

Figure 2.8
Soil and Water Program

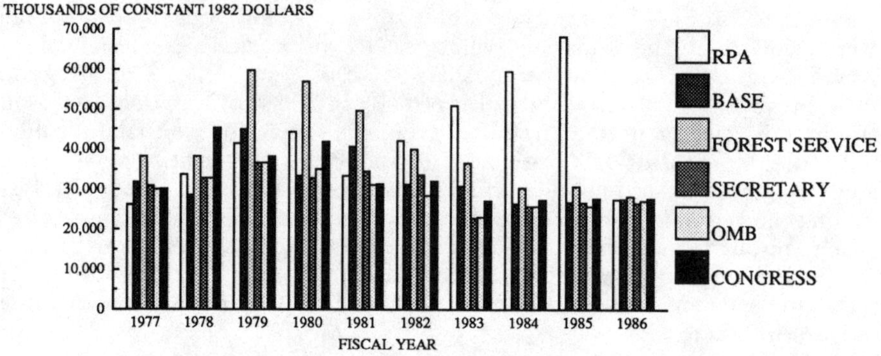

Figure 2.9
Minerals Program

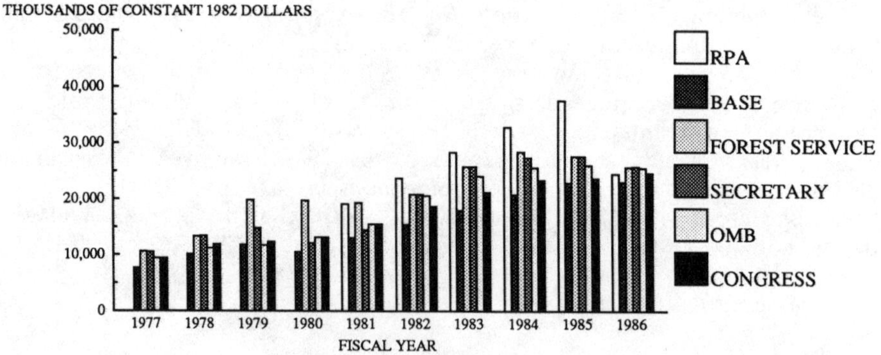

The Forest Service cooperates with the Department of Interior Bureau of Land Management, which is responsible for administering all energy and non-energy minerals on all federal lands, including the national forests. Before minerals can be removed, the Forest Service must complete all environmental analyses and issue lease permits or permits to operate a mining claim. The tasks in this program are primarily processing new lease applications and prospecting permits, and administering operating plans to minimize off-site impacts from mining activities.

For FY 1977-1981, the actions of the various participants in the budget process on the minerals program are similar to those on the other non-timber programs except that the response of the Congress was a very regular stairstep of increases that began in the mid-1970s and continued uninterrupted right through the change in administration and into the mid-1980s (see Figure 2.9). From FY 1977 through FY 1986, the budget for the minerals program increased by more than 163 percent, or more than two and a half times. Executive branch budget proposals for this program increased sharply with the advent of the Reagan administration and continued strong through FY 1986 when proposals for nearly every other resource program were being cut back. For FY 1982-1986, a time when funding for all other programs including timber were being scaled back, the minerals program enjoyed an increase of 58 percent. Most of the increased expenditures were to help the Forest Service meet the increasing number of applications for permits to explore for and develop leasable minerals on national forest lands. With an eye to the growing federal budget deficit, participants including Congress seemed well aware of this program's potential for producing net revenues to the Treasury.

Protection

The fire protection program in this analysis is largely limited to fuels management. The costs of actually fighting forest fires vary widely from one year to the next, depending on the incidence of fires and are usually provided for with a separate congressional appropriation once the fire season has ended.[6] The purpose of fuels management is to reduce the volume of naturally combustible material on the forest floor and thus minimize the potential for large, destructive wildfires. These activities include controlled (or "prescribed") burning, disposal of brush and logging debris and conducting surveys of fuel hazards.

Perhaps the boldest Forest Service initiative to be seen for any program in recent years was the agency's proposal to more than quadruple the funding for fire protection in FY 1979 (see Figure 2.10). Fire protection involves only "pre-suppression" activities such as brush disposal and does not include funding for actually fighting forest fires.[7] The secretary was clearly unimpressed, however, and actually recommended an 11 percent reduction from the previous year's funding. In a move that seems quite uncharacteristic, OMB then proposed to more than double the funding for these activities. Having come into office in January 1977, just as the FY 1978 President's Budget was already being submitted to Congress, the new Carter administration did not have full control of the preparation of this document until FY 1979. Differences in viewpoints between executive branch organizations are not uncommon, as will be discussed at greater length below, but it was unusual that OMB responded so positively to an agency's request for a sharp increase in funding and that is was the Office of the Secretary--rather than OMB--that attempted to apply the brakes.

The FY 1979 Forest Service initiative must be judged a success in that it resulted in actual appropriations more than double those of the previous year. Once this new plateau had become established, however, budget proposals for these activities relative to previous years' funding rapidly moderated. Congressional appropriations, and the various budget proposals leading up to them, have remained virtually constant in real terms right up through FY 1986. No such drama accompanied the change in administrations between the preparation of the FY 1981 budget and the preparation of the FY 1982 budget. No budget goals were established for these activities in the 1975 RPA Program and thus do not appear until FY 1981, the first year of the 1980 Program.

Lands

The lands program encompasses a variety of activities to facilitate the management of other resources on the national forests. The legal boundaries between national forest lands and other ownerships must be identified so that activities such as timber sales can be carried out without fear of trespass and to protect national forest lands from encroachment. Land exchanges and acquisitions are carried out primarily to reduce the cost or improve the effectiveness of resource management on the national forests. Exchanges consolidate national forest lands, making it more efficient to manage and administer various resource programs. In a recent year, 118,996 acres of non-federal land was acquired in exchange for 80,229 acres of national forest land, reducing property lines by more than 1,530 miles and saving about $8.5 million in future property line location costs.[8]

Although they seem dwarfed in Figure 2.11 by the magnitude of the budget goals in the 1980 RPA Program, the variations between budget proposals and between fiscal years for lands program activities (land acquisition, exchange, boundary location) are nonetheless significant. For both FY 1978 and FY 1979, the Forest Service proposed more than doubling the funding for these activities over previous years' appropriations. Despite a lack of support for this idea

Forest Service Budgets and RPA Program Targets
Fiscal Years 1977 - 1986

Figure 2.10
Protection

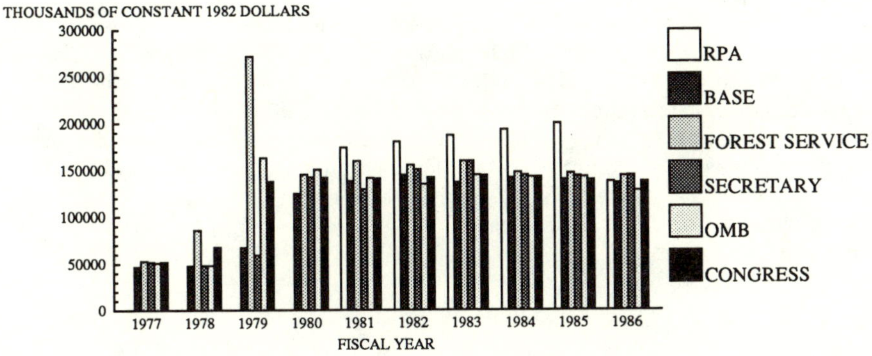

Figure 2.11
Lands Program

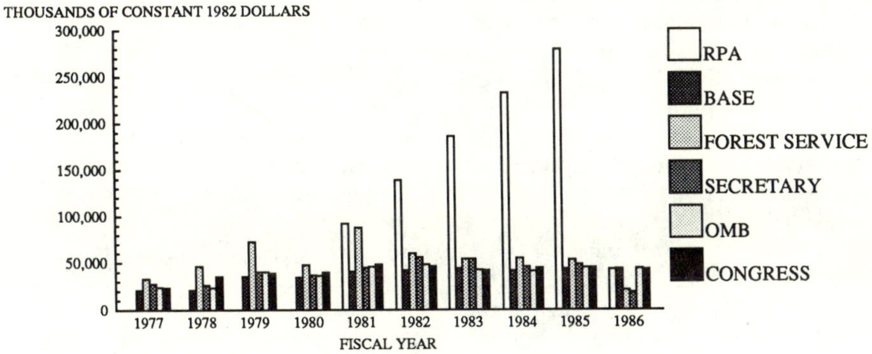

from the secretary or OMB, Congress responded positively for FY 1978 (providing a 67% increase over the previous year's funding) and for FY 1979 (a further 8% increase). Funding since that time has remained fairly constant, and the program has done relatively well to have received significant real-term increases in all but two years over the past decade. These activities seem to have suddenly fallen out of favor with the secretary in FY 1986, when his office proposed a 57 percent reduction in funding. The attitude of the secretary may have had some bearing on the similarly drastic 52 percent cut the Forest Service recommended for itself in that year. The secretary's view was apparently not shared in other parts of the Administration, however, and OMB recommended a continuation of existing funding (0.4% real-term increase over base). Congress seems to have agreed more with OMB, maintaining funding within 1 percent of the existing level. Throughout the period for which the 1980 RPA Program established budget targets, FY 1981-1986, these recommendations appear to have had little influence on either the proposals of the various participants or on final appropriations. Funding for these activities remained fairly constant for the period despite RPA recommended one-year increases of as much as 647 percent over the prior year's base.

Facilities

Due to the large staff and equipment required to manage the national forests, the Forest Service owns or leases more than 15 million square feet of building space and maintains more than 350,000 miles of forest access roads.[9] Constructed with a life expectancy of 30 to 35 years many of the buildings are functionally obsolete, posing rising maintenance costs. Historically, funding for facility maintenance has been less than one percent of replacement value, and this has resulted in a rising backlog of postponed maintenance tasks.[10] Road maintenance is financed primarily by appropriations but also gets substantial support from requirements on commercial users such as timber purchasers. At any given time, only about 80 percent of the road system is open for public use, and only about 30 percent of the system is usable by vehicles without high clearance and/or four-wheel drive.

Facilities construction and maintenance was a major initiative of the 1980 RPA Program. In spite of the Forest Service's unsuccessful attempts to up the ante during the late 1970s and early 1980s, however, actual funding for this program remained largely unchanged in real terms for the ten years, FY 1977-1986. Administration officials were passively supportive of funding increases in the first years of both the Carter and Reagan administrations, but in both cases this support soon waned. The 1980 RPA Program recommended steady increases for FY 1981-1985, up to two and half times the prevailing base. In contrast, the annual growth in appropriations generally declined from

FY 1977 onward with reductions from the prior year base (in real terms) for every year after FY 1983.

Differences in Program Emphasis Among Budget Process Participants

The relative emphasis that each participant in the budget process places on individual resource programs within the National Forest System budget reveals a great deal about that participant's organizational values. The allocation of budgetary resources among different programs affects many different interests, public and private, and is thus an inherently political process. Anyone at all familiar with the federal budget process would certainly expect to see differences in program emphasis between the executive branch budget proposal and subsequent congressional appropriations. What may not be expected, and not easily discernible given the limitations on public disclosure discussed above, are differences in program emphasis within the executive branch itself.

To the extent that data were available, the budget proposals of each participant were compared against one another as percentage changes from a common basis, the prior year's actual appropriation. Among the resource programs during fiscal years 1977-1986, the greatest average annual increases proposed by the Forest Service were, in decreasing rank, for the fish and wildlife program (55%), followed by those for the minerals program (42%), recreation (31%), soil and water (26%), range (24%) and finally timber (17%) (see Table 2.1). The ranking differs somewhat for USDA, from minerals to fish and wildlife, timber, range, recreation, then soil and water. The ranking differs yet again for OMB, where the order is: minerals, timber, fish and wildlife, recreation, soil and water, then range. A second and closer examination of the data for FY 1977-1986 indicates that these averages could be somewhat misleading in that, for some participants, there seemed to be two clearly different sets of priorities within the ten-year period. A shift point can be seen between FY 1981 and FY 1982, corresponding to the change in presidential administrations (see Table 2.3). Perhaps the most pronounced shift was in the proposals from the Forest Service itself in relation to those from the secretary of agriculture. During the first period, FY 1977-1981, the Forest Service proposed increases over the base for the timber program that averaged 23 percent per year (see Table 2.3) and much greater increases for all other programs, ranging from 36 percent for the soil and water program to 88 percent for the fish and wildlife program. During the second period, FY 1982-1986, the Forest Service significantly scaled back its requests for all the non-commodity programs, both in absolute terms and in relation to its requests for the timber program. The secretary's office, meanwhile, more than doubled its average proposed increase for the timber program from 5 percent in period one to 11 percent in period two.

Table 2.3
Average Percent Change of Budget Proposals from the Prior Year Base, by Participant and Resource Program, Showing Shift from Period 1 to Period 2

PROGRAM	RPA	USFS	USDA	OMB	HOUSE	SENATE	APPROPRI -ATIONS
			[PERCENT]				
PERIOD 1: FY 1977-1981							
TIMBER	37.8	22.5	5.2	5.8	15.6	14.7	15.3
RECREATION	132.8	55.6	1.3	-0.4	15.3	9.3	13.1
WILDLIFE & FISH	80.7	87.5	14.3	7.6	37.4	28.8	30.4
RANGE	73.9	38.7	2.3	-2.7	15.1	10.2	11.0
SOIL & WATER	1.6	35.7	-4.5	-5.4	10.8	4.2	8.2
MINERALS	*	55.0	24.4	15.6	18.0	16.6	17.9
PROTECTION	*	85.8	1.8	35.1	15.6	32.5	34.7
LANDS	*	87.1	17.3	10.9	32.6	20.4	24.2
FACILITIES	*	51.4	6.0	3.1	13.9	9.1	12.3
NFS RESOURCE PROGRAMS - TOTAL	*	35.2	4.9	6.1	15.6	14.0	15.5
PERIOD 2: FY 1982-1986							
TIMBER	1.7	11.0	11.1	0.3	-10.2	-3.3	-3.5
RECREATION	72.0	6.8	-3.9	-13.3	-4.4	-11.2	-2.5
WILDLIFE & FISH	91.5	19.9	7.4	-2.3	-0.2	2.7	0.1
RANGE	48.7	10.0	0.4	-6.0	-7.4	1.6	-3.6
SOIL & WATER	75.4	15.4	-5.0	-8.4	-0.0	-4.7	-0.6
MINERALS	47.3	29.1	28.2	22.8	12.1	21.2	12.2
PROTECTION	28.0	7.7	6.2	-0.9	1.0	2.1	1.0
LANDS	312.0	13.7	4.1	2.6	3.4	2.7	2.8
FACILITIES	63.1	22.1	10.2	-1.4	-4.1	-1.9	-3.1
NFS RESOURCE PROGRAMS - TOTAL	29.0	11.8	8.6	-1.2	-7.1	-2.8	-2.6
DIFFERENCE FROM PERIOD 1 TO PERIOD 2:							
TIMBER	-36.1	-11.5	5.9	-5.5	-25.8	-18.0	-18.8
RECREATION	-60.8	-48.8	-5.2	-12.9	-19.8	-20.6	-15.6
WILDLIFE & FISH	10.9	-67.7	-6.9	-10.0	-37.7	-26.0	-30.3
RANGE	-25.2	-28.6	-1.9	-3.3	-22.5	-8.6	-14.6
SOIL & WATER	73.9	-20.3	-0.4	-3.0	-10.9	-9.0	-8.7
MINERALS	47.3	-25.9	3.8	7.2	-6.0	4.6	-5.7
PROTECTION	28.0	-78.1	4.5	-36.0	-14.6	-30.4	-33.6
LANDS	312.0	-73.5	-13.2	-8.3	-29.2	-17.7	-21.5
FACILITIES	63.1	-29.4	4.3	-4.5	-18.0	-11.1	-15.3
NFS RESOURCE PROGRAMS - TOTAL	29.0	-23.4	3.8	-7.3	-22.7	-16.8	-18.1

Moreover, there was a change in the relationship between the Forest Service and the secretary's office. During period two, the secretary's office continued to restrain the Forest Service requests for nearly all programs. For the timber program, however, the secretary's office went from restraining the Forest Service during period one to actually increasing the agency's request during period two, a situation unique to timber among all the National Forest System resource programs. This anomaly becomes all the more puzzling considering that (1) period one was a time of strong market demand for timber and period two witnessed the deepest timber industry recession in several decades, and (2) the secretary's attempts to increase funding for the timber program came on top of Forest Service requests in which the timber program constituted 6 percent *more* in period two than it had in period one, the recession in the timber market notwithstanding (see Table 2.4).

Among the other budget process participants, the shifts from period one to period two seemed to be more even-handed reductions aimed simply at getting control of the looming federal budget deficits of the time. OMB permitted slight increases in the NFS lands program, held the timber program basically constant and requested moderate reductions in all other programs. The exception to this was a continued strong emphasis on the revenue-generating NFS minerals program, for which OMB still proposed an average annual 23 percent increase. The NFS minerals program was also the anomaly in the House and Senate proposals, where it continued to receive strong support while all other programs were being gradually cut back. The emphasis on mineral development on the national forests, especially for oil and gas, reflected the shifts in national energy policy in response to the energy shortages of the mid-1970s. In the House and Senate, the differences in proposals from period one to period two appear to be fairly consistent double-digit reductions across all NFS resource programs except minerals. As the budget tide went out, all boats seemed to have been allowed to settle more or less equally. This is especially well illustrated in Table 2.4, showing the distribution of funding among the different resource programs in each period for each budget process participant. Despite period two's double-digit reductions by the House and Senate, the *distribution* of funding among programs was virtually identical to that proposed by the House and Senate during period one. For the most part, the proportion of total NFS funding constituted by each resource program changed by no more than approximately 1 percent. The necessity of budget cutbacks brought overall budget reductions but no significant shift in budget emphasis among the resource programs.

This was less the case with the executive branch where there was a more perceptible shift from non-commodity programs to a greater emphasis on commodity programs. In terms of the differences between the distributions from period one to period two, the greatest shift in this direction was in the Forest Service requests, followed

Table 2.4
Distribution of Funding Among NFS Resource Programs, by Participant, FY 1977-1986

RESOURCE PROGRAM	RPA	BASE	USFS	USDA	OMB	HOUSE	SENATE	APPROPRI-ATIONS
			[PERCENT]					
PERIOD 1: FY 1977-1981								
TIMBER	71.0	77.2	73.0	77.5	77.9	77.1	77.8	77.5
RECREATION	17.8	11.6	14.1	11.3	11.2	11.5	11.2	11.4
FISH & WILDLIFE	3.0	2.5	3.5	2.7	2.6	2.9	2.8	2.8
RANGE	4.0	3.4	3.7	3.4	3.2	3.4	3.4	3.3
SOIL & WATER	2.7	4.1	4.2	3.7	3.7	3.8	3.6	3.7
MINERALS	1.5	1.2	1.4	1.4	1.3	1.2	1.2	1.2
NFS RESOURCE PROGRAMS-TOTAL	100.0	100.0	100.0	100.0	100.0	100.0	100.0	100.0
PERIOD 2: FY 1982-1986								
TIMBER	70.4	79.6	79.5	81.2	80.8	77.9	79.3	79.1
RECREATION	15.5	10.5	10.1	9.2	9.2	11.1	9.8	10.6
FISH & WILDLIFE	4.6	2.8	3.1	2.8	2.8	3.2	3.1	3.0
RANGE	3.5	2.7	2.7	2.5	2.6	2.8	2.9	2.7
SOIL & WATER	3.8	2.6	2.7	2.2	2.4	2.8	2.6	2.6
MINERALS	2.3	1.8	2.1	2.1	2.2	2.2	2.3	2.1
NFS RESOURCE PROGRAMS-TOTAL	100.0	100.0	100.0	100.0	100.0	100.0	100.0	100.0
SHIFT IN BUDGET EMPHASIS, PERIOD 1 TO PERIOD 2:								
TIMBER	-0.7	2.3	6.5	3.7	2.9	0.8	1.6	1.6
RECREATION	-2.3	-1.0	-4.1	-2.1	-2.0	-0.4	-1.3	-0.8
FISH & WILDLIFE	1.7	0.3	-0.5	0.1	0.2	0.2	0.3	0.1
RANGE	-0.5	-0.7	-1.0	-0.9	-0.7	-0.7	-0.5	-0.7
SOIL & WATER	1.1	-1.5	-1.5	-1.4	-1.3	-0.9	-1.1	-1.1
MINERALS	0.8	0.6	0.6	0.7	0.9	1.0	1.0	0.8

by those from the secretary's office and then OMB. Despite the relatively small percentage changes, the reallocations of budget resources that they represent are not insignificant. For recreation to represent 4 percent less of the overall budget in period two translates to a more than 29 percent reduction in the actual funding of the program, assuming overall budget levels remain constant; the fact that the overall NFS budget fell significantly in real terms during period two only accentuates the impact on programs receiving relatively less emphasis.

Perhaps the easiest way of seeing how the priorities and desired resource program emphasis varied among the different budget process participants is to visually compare the budget emphasis for each year by each participant. Figures 2.12 to 2.17 show the recommended increase or decrease from the prior year's appropriations for each resource program in a way that allows the differing program emphasis between the participants to be easily recognized. For example, Figure 2.12 clearly shows that the Forest Service's priority for increasing timber program funding was very low relative to the other resource programs from FY 1978 through FY 1982. At the secretary's office, timber was an even lower priority during the late 1970s, then dramatically took precedence beginning in FY 1982 (see Figure 2.13); the minerals program alone remained a consistent high priority, with the recreation, and soil and water programs singled out for relatively large reductions through the 1980s. OMB's priorities appeared to be reasonably consistent with those of the secretary's office during the late 1970s, but departed sharply from the secretary's views in FY 1982 and beyond, especially regarding increases in the timber program (see Figure 2.14). The House seemed to generally favor increasing in the non-commodity resource programs and maintaining the timber program very close to the status quo until FY 1984 (see Figure 2.15); like OMB it could not ignore the condition of the timber markets by providing the sort of increases for the timber programadvocated by the secretary, and its concern over the below-cost timber sales issue became clearly evident in its actions for FY 1986. Figures 2.16 and 2.17 show that although there were no easily discernible shifts in program emphasis from the House to the Senate to the final conference, the amplitude of the deviations from the prior year base seem decrease at each stage, especially during the mid-1980s. As the budget progresses from agency request to conference bill, there seems to be less and less flexibility to make more than small incremental changes (except in the case of a major policy dispute, such as over below-cost timber sales and the timber program in FY 1986).

Correlations of Budget Proposals with Final Appropriations

A comparison of the budget proposal of each participant with what was ultimately appropriated is useful in assessing the effectiveness

BUDGETS PROPOSED FOR THE FOREST SERVICE
Percent Change from Prior Year Appropriations, by Resource Program, FY 1977-1986

Figure 2.12
Forest Service Request

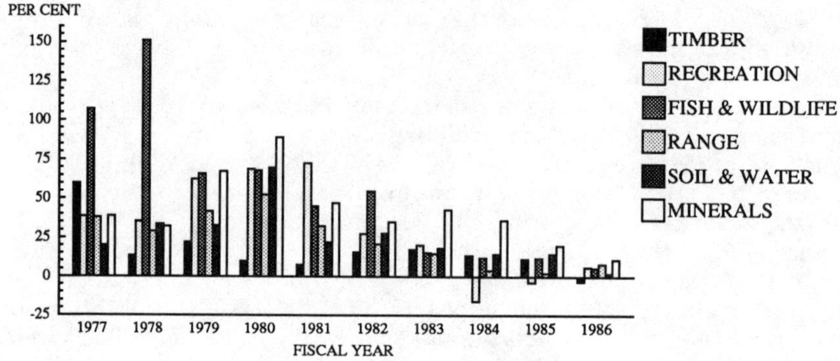

Figure 2.13
Secretary's Allowance

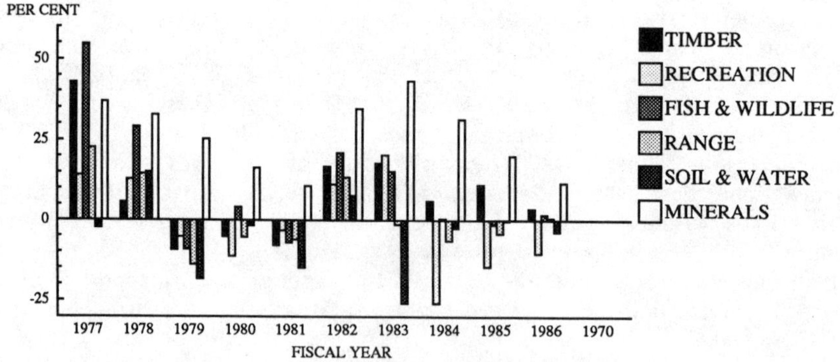

BUDGETS PROPOSED FOR THE FOREST SERVICE
Percent Change from Prior Year Appropriations, by Resource Program, FY 1977-1986

Figure 2.14
President's Budget (OMB)

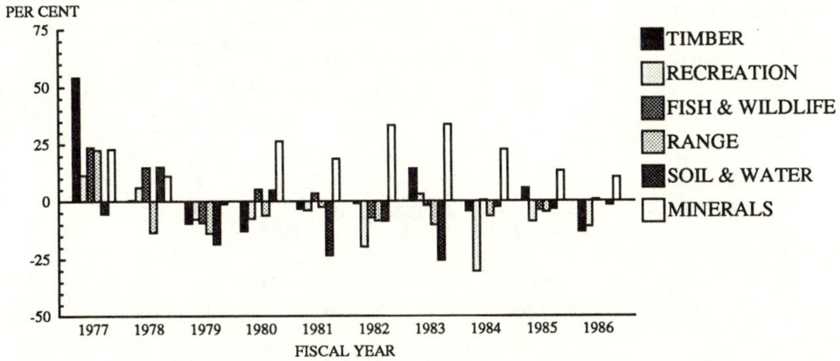

Figure 2.15
House of Representatives

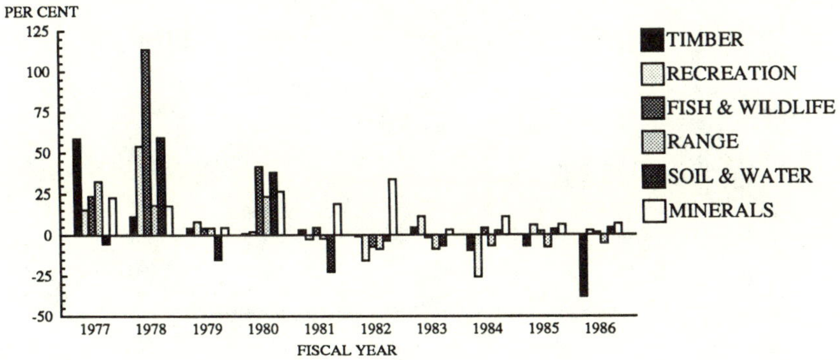

BUDGETS PROPOSED FOR THE FOREST SERVICE
Percent Change from Prior Year Appropriations, by Resource Program, FY 1977-1986

Figure 2.16
Senate

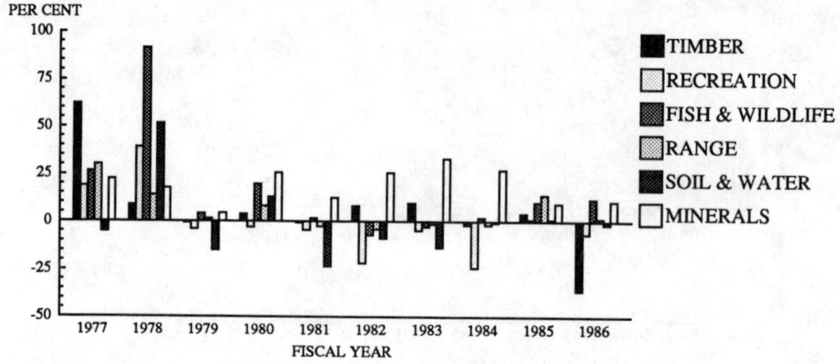

Figure 2.17
Congress

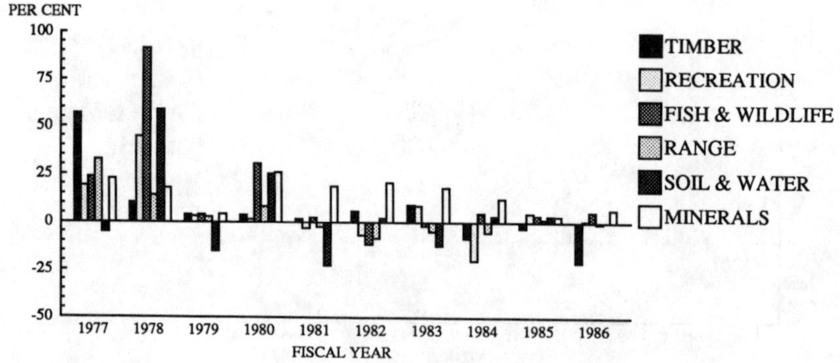

of each participant in influencing actual funding.[11] Table 2.5 shows the correlation, by resource program, between final appropriations and the budget proposal of each participant. For FY 1977-1986 OMB's budget proposals for the National Forest System consistently had a higher correlation with appropriations than that of either the Forest Service or the secretary's office. Moreover, for most of the National Forest System programs, OMB's correlation is higher than that of the prior year base, which is widely held to be the best single predictor of an agency's appropriations.[12] There are a few important exceptions to this general observation. One is the timber program, for which final appropriations are somewhat more consistent with the Forest Service request (r = .8845) than with either OMB (r = .8123) or USDA (r = .7033) proposals. The Forest Service also comes in closest on the range program (r = .9426), as compared to OMB (r = .6783) and USDA (r = .5306). The rankings and percentages are comparable for construction and maintenance of facilities. The correlation of the prior year base with appropriations is actually quite low for all resource programs except minerals. Correlation analysis of the longer FY 1960-1986 time-series also shows a stronger relationship between appropriations and OMB than with the prior year base, across all National Forest System programs except the range program.

The Impact of RPA on Forest Service Program Funding

The real-term growth in funding for National Forest System programs from FY 1977 through FY 1981 seems truly phenomenal. Seldom has any federal agency experienced funding increases as large and sustained (see Table 2.6). To many observers inside and outside the Forest Service, this unprecedented growth in program funding seemed to be a direct result of RPA planning and the budget targets established for each resource program in the RPA Program documents. The RPA Assessment and Program not only provided a more thorough and comprehensive foundation on which to justify annual budget requests, but also served as a uniquely valuable device for focusing the attention of the critical authorizing and appropriations subcommittees in the Congress, if only for a moment, solely on the needs of the Forest Service. Even in the subcommittee that determines its appropriations, the Forest Service generally must compete for attention with everything from natural gas leasing on the Outer Continental Shelf to the granting policies of the National Endowment for the Arts. Any document that can bring one particular agency the undivided consideration of key congressional legislators is a valuable tool indeed. Several other agencies soon began evaluating the possibility of seeking legislation that could do for them what RPA had done for the Forest Service.[13] Figures 2.18 to 2.24 show that during the five years following the first RPA Program, the annual rate of budget growth for virtually every program far exceeded

Table 2.5
Correlation of Proposed Changes from the Prior Year Base with Changes in Actual Appropriations, by Participant and Resource Program

DEPENDENT VARIABLE: APPROPRIATIONS

INDEPENDENT VARIABLE:	NFS - TOTAL	TIMBER	RECREATION & WDNS	FISH & WILDLIFE	RANGE	SOIL & WATER	MINERALS	PROTECTION	LANDS	FACILITIES
FISCAL YEARS 1977-1986:										
RPA (HIGH)	0.7805	0.6490	0.6813	0.8120	0.6056	0.2919	0.7154	0.6539	0.1863	0.1334
RPA (LOW)	(1)	0.9493	0.6563	0.6669	0.8648	0.5740	0.8100	(1)	(1)	(1)
BASE	0.8246	0.6556	0.4899	0.7645	0.7266	0.3106	0.9855	0.8482	0.8654	0.5757
USFS	0.8819	0.8845	0.4636	0.9426	0.5100	0.6190	0.9256	0.7157	0.4778	0.8282
USDA	0.7584	0.7033	0.3801	0.5306	0.5283	0.6759	0.9719	0.8047	0.5528	0.4288
OMB	0.8617	0.8123	0.6867	0.6783	0.7838	0.8284	0.9934	0.9650	0.8880	0.7019
HOUSE	0.9211	0.8909	0.9454	0.9647	0.9677	0.9774	0.9794	0.8243	0.9448	0.9610
SENATE	0.9271	0.9156	0.8778	0.9547	0.9664	0.9772	0.9915	0.9984	0.9721	0.9395
FISCAL YEARS 1960-1986:										
BASE	0.9743	0.9657	0.9246	0.9461	0.9613	0.8731	0.9754	0.9420	0.9790	0.8732
OMB	0.9896	0.9863	0.9671	0.9432	0.9748	0.9767	0.9922	0.9880	0.9820	0.9696
HOUSE	0.9964	0.9938	0.9919	0.9824	0.9970	0.9977	0.9919	0.9490	0.9948	0.9924
SENATE	0.9956	0.9924	0.9887	0.9899	0.9967	0.9948	0.9946	0.9942	0.9971	0.9950

(1) Budget goals for these resources not established in RPA Programs prior to 1980.

Table 2.6
Annual Percent Increase in Funding for National Forest Resource Programs, As Compared with the Federal Budget

FISCAL YEAR	FEDERAL BUDGET (1)	NATIONAL FOREST SYSTEM	TIMBER	RECREATION	WILDLIFE & FISH	RANGE	SOIL & WATER	MINERALS	FIRE	LANDS	FACILITIES
	[PERCENT, REAL TERMS]										
1977	6.87	37.33	52.84	14.83	19.69	27.77	-8.17	32.79	6.14	11.10	21.03
1978	15.06	25.80	19.65	43.56	91.47	37.89	51.25	14.56	27.17	51.88	27.04
1979	-3.65	10.73	0.89	7.28	5.85	0.51	-16.67	2.66	176.21	6.46	1.11
1980	4.29	-3.82	-7.62	-9.23	12.78	-3.46	9.70	6.29	11.13	4.49	-7.84
1981	-5.46	0.27	1.94	-5.19	-6.65	-14.92	-24.20	19.52	4.66	6.50	-2.44
1982	-11.82	4.71	12.19	-6.94	-1.25	4.42	-1.52	15.78	-11.94	4.03	-0.28
1983	-5.91	6.04	11.48	32.51	-3.04	-6.62	-13.60	16.48	-29.36	-5.09	15.98
1984	-3.20	-9.84	-12.61	-22.59	1.62	-8.64	-0.01	8.87	18.83	2.64	-18.49
1985	13.74	2.82	2.33	1.84	0.70	-0.44	2.95	0.36	10.48	0.52	-1.64
1986	-20.59	-11.26	-21.67	-5.78	-1.91	-7.73	-6.79	-0.71	41.30	-61.13	-9.58
AVG 1977-1986	-1.07	6.28	5.94	5.03	11.93	2.88	-0.71	11.66	25.46	2.14	2.49
AVG 1977-1981	3.42	14.06	13.54	10.25	24.63	9.56	2.38	15.16	45.06	16.09	7.78
AVG 1982-1986	-5.56	-1.51	-1.65	-0.19	-0.78	-3.80	-3.79	8.16	5.86	-11.81	-2.80
AVG 1960-1986	4.73	5.73	7.55	6.32	11.42	2.82	11.46	5.14	13.59	5.27	3.13

(1) Net of defense and entitlement programs.

that for the federal government overall. This is especially true for many of the non-commodity resource programs such as fish and wildlife, soil and water, and recreation, for which appropriations nearly doubled in just a few years, often after a decade or more of static, low-level funding. But a broader perspective on federal spending suggests that this phenomenon may be due, at least in part, to a general emphasis during this period, by both the executive branch and the Congress, on natural resource conservation and environmental protection. Figure 2.25 shows annual appropriations for the natural resources and environment (NR&E) function of the federal budget. The NR&E function includes all renewable resources, water, and pollution control programs in the departments of Agriculture and Interior as well as the Environmental Protection Agency and resource programs of the Army Corps of Engineers. In order to make the data more directly comparable to that for the Forest Service, funding for water resources development, OCS leasing and pollution control programs have been factored out of the Natural Resources and Environment funding represented in Figure 2.25. The responsibilities of the Department of Interior, Bureau of Land Management (BLM) may correspond even more directly to those of the Forest Service in that BLM also manages large areas of federal lands for many of the same resources found on the national forests. Figure 2.26 also compares annual appropriations for BLM and the Forest Service.

For FY 1977-1981, the period when the 1975 RPA Program was in effect, the average annual real increase in funding for National Forest System programs was more than 14 percent (see Table 2.6). During this same period, annual real growth for the overall federal budget, net of defense and entitlement programs, averaged only 3.4 percent. When only the NR&E function is considered, this annual growth rate rises to 4.7 percent, still significantly below that of Forest Service programs (see Table 2.7). A direct comparison with BLM, however, shows average annual growth rates and patterns over time quite comparable to those for National Forest System programs. Funding for BLM renewable resource programs grew at an average annual rate of 12.2 percent.[14] It could be argued that this figure is most appropriately compared to funding for the Forest Service overall, for which the average annual growth in funding, 20.3 percent, is considerably higher than BLM. However, overall Forest Service funding includes that for a large research organization, for which thereis no real BLM counterpart, and for state and private forestry programs, largely a "pass through" program to channel federal funds to state forestry organizations to encourage forest management on non-federal lands. A comparison of BLM to the National Forest System, for which the average annual growth rates during this period are less than two percentage points apart, concentrates on resource management programs by the two federal agencies on federally owned lands.

Figure 2.18
Relative Growth of the Forest Service Budget,
Fiscal Years 1960-1986

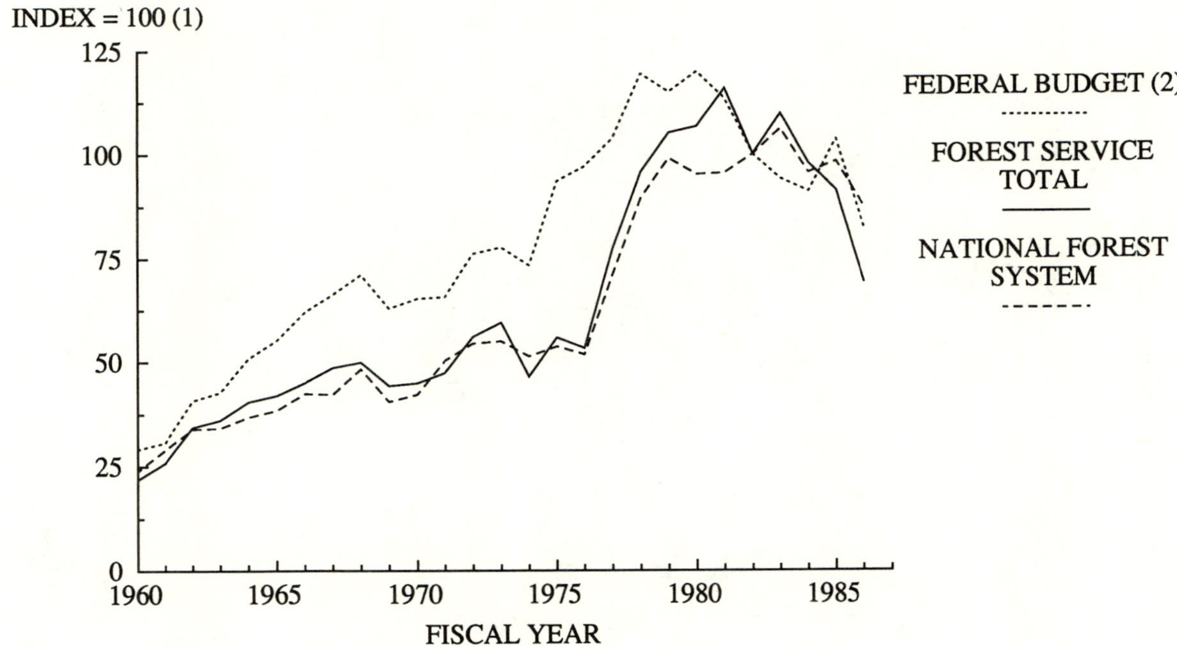

(1) Index based on constant 1982 dollars
(2) Net of defense and entitlement programs

Relative Growth of the Forest Service Budget, Fiscal Years 1960-1986

Figure 2.19
Timber Program

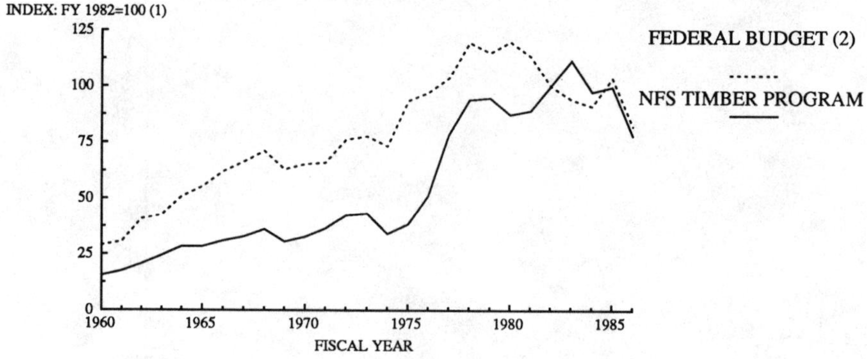

Figure 2.20
Recreation Program

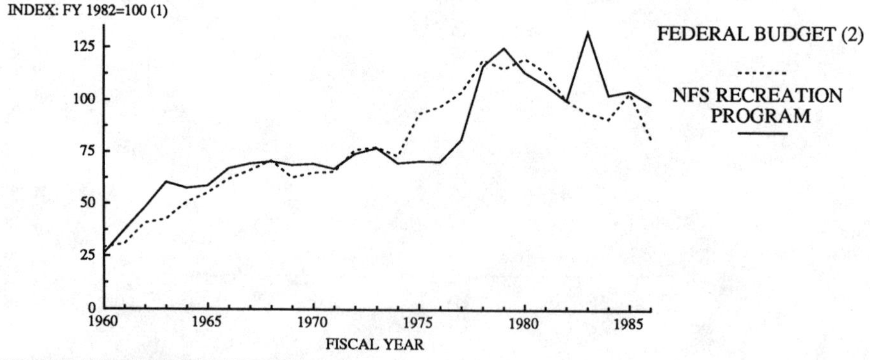

(1) INDEX BASED ON CONSTANT 1982 DOLLARS
(2) NET OF DEFENSE AND ENTITLEMENT PROGRAMS

Relative Growth of the Forest Service Budget, Fiscal Years 1960-1986

Figure 2.21
Fish and Wildlife Program

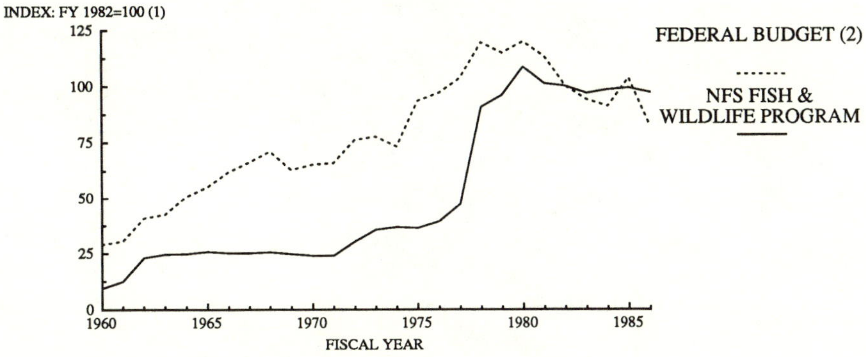

Figure 2.22
Range Program

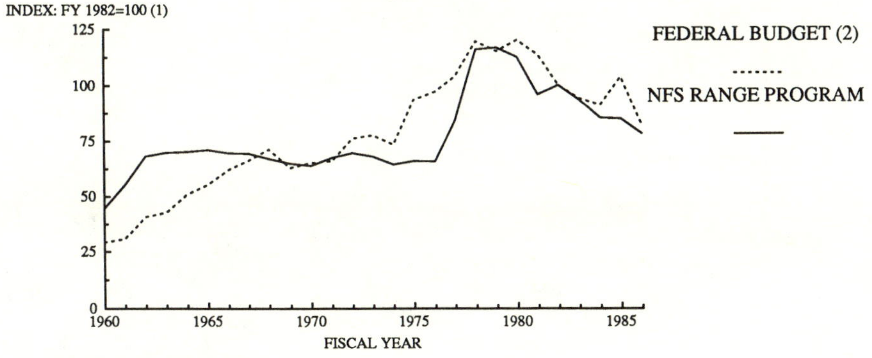

(1) INDEX BASED ON CONSTANT 1982 DOLLARS
(2) NET OF DEFENSE AND ENTITLEMENT PROGRAMS

Relative Growth of the Forest Service Budget, Fiscal Years 1960-1986

Figure 2.23
Soil and Water Program

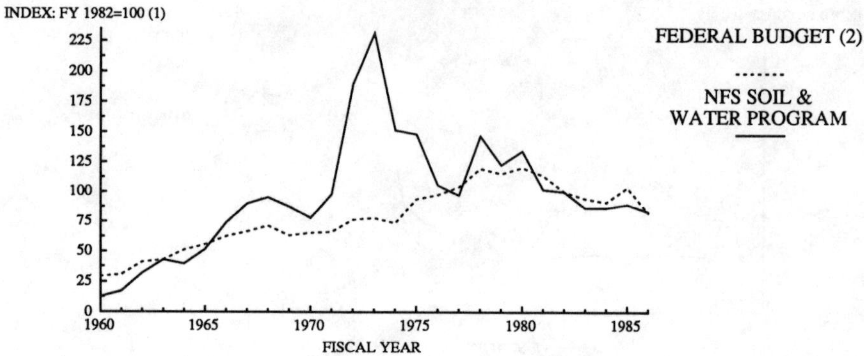

Figure 2.24
Minerals Program

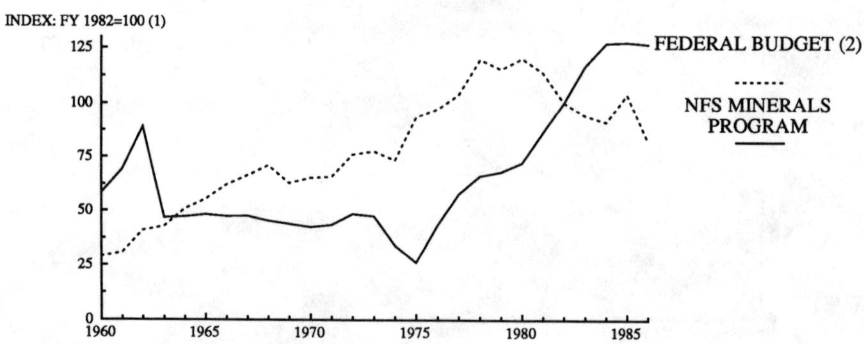

(1) INDEX BASED ON CONSTANT 1982 DOLLARS
(2) NET OF DEFENSE AND ENTITLEMENT PROGRAMS

Relative Growth of the Forest Service Budget, Fiscal Years 1960-1986

Figure 2.25
Natural Resources and Environment Portion of the Federal Budget

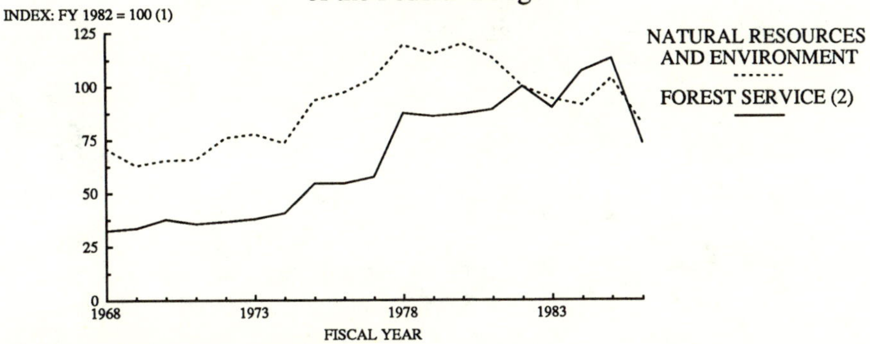

(1) INDEX BASED ON CONSTANT 1982 DOLLARS
(2) STATE REVENUE SHARING PROGRAMS NOT INCLUDED

Figure 2.26
Bureau of Land Management (Department of Interior)

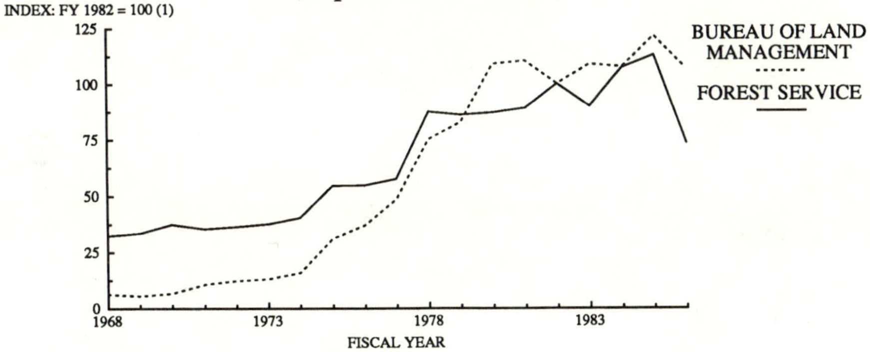

(1) INDEX BASED ON CONSTANT 1982 DOLLARS
(2) STATE REVENUE SHARING PROGRAMS NOT INCLUDED

Table 2.7
Annual Percent Increase in Funding in National Forest Programs, As Compared with Other Federal Natural Resources Programs

FISCAL YEAR	FEDERAL BUDGET (3)	ALL NATURAL RESOURCES & ENVIRONMENT FUNCTIONS (NR&E)	NR&E FUNCTIONS NET OF WATER RESOURCES & POLLUTION ABATEMENT & CONTROL	BLM (1)	FOREST SERVICE (2)	NATIONAL FOREST SYSTEM ONLY
	[PERCENT, REAL TERMS]					
1977	6.87	14.94	0.32	64.87	56.37	37.33
1978	15.06	2.06	31.67	3.18	17.52	25.80
1979	-3.65	1.47	-8.50	-0.28	12.23	10.73
1980	4.29	4.75	5.85	-5.86	0.55	-3.82
1981	-5.46	-10.74	-5.71	-0.79	14.57	0.27
1982	-11.82	-9.95	-11.02	-5.35	-14.91	4.71
1983	-5.91	-6.08	7.48	-8.85	17.47	6.04
1984	-3.20	-4.76	-4.73	17.63	-14.16	-9.84
1985	13.74	2.85	3.28	21.68	-9.00	2.82
1986	-20.59	-6.16	-11.63	-32.55	-26.51	-11.26
AVG 1977-1986	-1.07	-1.16	0.70	5.37	5.41	6.28
AVG 1977-1981	3.42	2.50	4.73	12.22	20.25	14.06
AVG 1982-1986	-5.56	-4.82	-3.32	-1.49	-9.42	-1.51
AVG 1960-1986	4.73	2.66	3.08	6.91	6.63	5.73

(1) Figures for BLM are net of offshore minerals program now under Minerals Management S also net of state revinue sharing programs (Payments in Lieu of Taxes).
(2) Net of state revenue sharing programs (PILT and 25% Payments); includes funding for Research and State and Private Forestry activities.
(3) Net of defense and entitlement programs.

A 1980 report analyzing whether BLM should pursue legislation authorizing the preparation of a counterpart to the RPA Assessment and Program, concluded that factors other than RPA were largely responsible for the increases in the Forest Service budget. One major factor was the membership on the House Interior Appropriations Subcommittee of Congressman Bob Duncan from Oregon, who for several years aggressively sought higher funding levels for timber management on the national forests.[15] Nonetheless, it seems clear that RPA did play some role in the sudden, large and unprecedented funding increases that occurred just after the first RPA Program was produced by the Forest Service and presented to Congress. In particular, funding for the non-commodity resource programs such as fish and wildlife, soil and water, and recreation may have been greatly enhanced by RPA program planning. The 1975 RPA Assessment was the most comprehensive appraisal to date of the rapidly increasing demand for these resources on the national forests. And together with the Program, this represented the most thorough and persuasive argument yet presented to Congress to justify significant increases in funding for these programs. The dramatic upswings in the funding trends for these resource programs concurrent with the 1975 Program, as shown in Figures 2.19 to 2.24, cannot be easily dismissed.

Perhaps more importantly for the longer run, the rapid succession of funding increases for these non-commodity programs during the late 1970s helped establish a new plateau, without which the overall reductions in federal spending during the 1980s may have proved devastating to them. While this future was certainly not known to strategists and planners at the Forest Service in 1975, the history of economic cycles foretold that the coming of an era of budgetary cutbacks was not a question of "if" but "when." As with any new political entity that has the impetus of strong bipartisan support in Congress, there is a "window of opportunity" that can be either boldly exploited or lost through inaction. Judging by the Forest Service's very aggressive proposals for increased program funding during the late 1970s, this lesson was not lost on them and, judging by the resulting growth in appropriations, the strategy seems to have proved in some sense successful, despite its failure to bring about the recommended shift in program emphasis. A look at the events of the 1980s, however, suggests that this window of opportunity has all but closed, and that the greatest budgetary contributions of RPA planning may now be a matter of history.

Relationship of Budget Targets in the 1980 RPA Program with Actual Appropriations

In Table 2.5, the correlation between actual appropriations and the budget goals established in the three RPA Programs is provided for reference. These correlations are quite low for most NFS programs, in most cases ranking below those of the prior year base and all of the various participants in the budget process. It will be recalled that there were two alternative Programs presented to Congress in 1980. One, the "high-bound" alternative, was produced independently by the Forest Service and USDA; the second, or "low-bound" alternative, was developed largely by OMB. With an air of righteous indignation, Congress at the time roundly rejected the low-bound alternative as yet another instance of excessive meddling by OMB, and endorsed the high-bound alternative as the new 1980 RPA Program. Figures 2.27 through 2.36 show, for each resource program, the budget targets for FY 1981-1985 from the high-bound alternative as compared with actual appropriations for those years. Also displayed are the budget targets for each program that were established in the low bound alternative.

For the timber program, actual appropriations follow the funding targets from the high-bound alternative reasonably well. By the end of the period covered by the 1980 Program, FY 1985, appropriations for the timber program were at 86 percent of the RPA budget goal, despite setbacks due to the recession in the timber industry. For all the other resource programs, actual appropriations seemed to have little in common with the high-bound funding targets. By FY 1985, appropriations were at 44 percent of the RPA budget goal for recreation, 39 percent for fish and wildlife, and 41 percent for soil and water protection. Congressional appropriations departed significantly from not only the overall funding levels recommended in the Program, understandable during a period of general cutbacks in federal spending, but from the recommended *balance* of funding among the various resource programs as well.

Although congressional appropriations showed little congruency with the high-bound budget goals in the Program, they did bear a striking resemblance to the budget goals in the low bound alternative, which, though they had been ostensibly abandoned, still could be found in the 1980 Program document. In the correlations discussed above, actual appropriations were shown to have a stronger relationship to the OMB budget proposals than to those of the Forest Service or USDA, except in the case of the timber program. The correlation between actual appropriations and OMB's low-bound RPA alternative is even closer, to the point that the low-bound budget goals for the timber program correlate more closely with actual appropriations than even the House and Senate appropriations committee proposals (see Figure 2.27 and Table 2.5). That the actual appropriations for FY 1981-1985 follow very closely, for every resource program, with the RPA Program alternative prepared in 1980 by OMB hardly seems coincidental. Though Congress may have explicitly rejected OMB's longer-term views on what resource management activities should take place on the national forests, it in effect heeled quite closely to those views in its subsequent budget decisions. What OMB could not accomplish through overt modifications to Forest Service program planning it simply accomplished through the annual budget process.

RPA Alternative Budget Targets and Actual Appropriations, by Resource Program, Fiscal Years 1977-1986

Figure 2.27
Timber Program

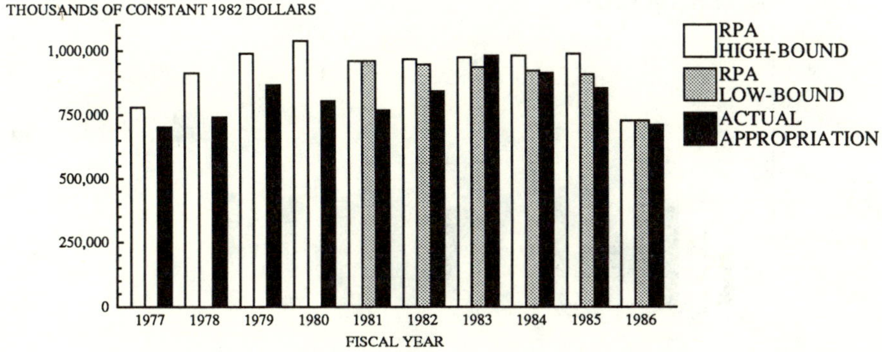

Figure 2.28
Recreation Program

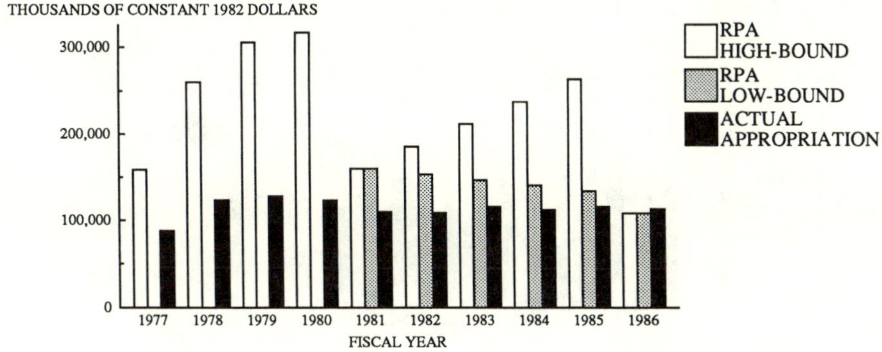

RPA Alternative Budget Targets and Actual Appropriations, by Resource Program, Fiscal Years 1977-1986

Figure 2.29
Fish and Wildlife Program

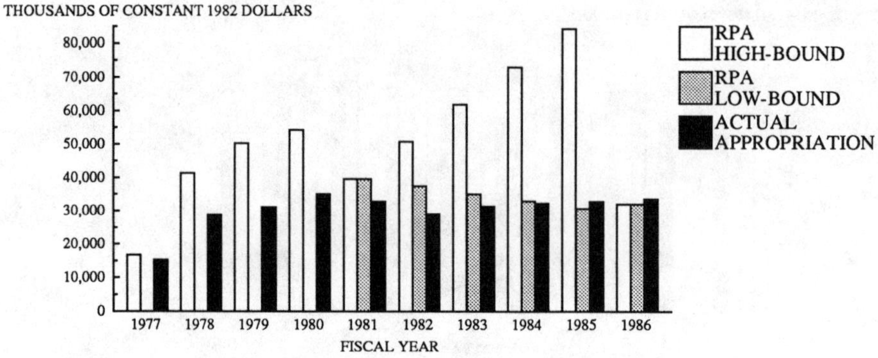

Figure 2.30
Range Program

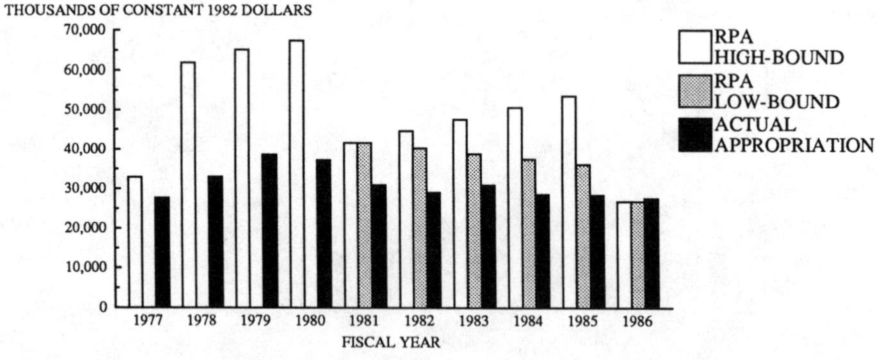

RPA Alternative Budget Targets and Actual Appropriations, by Resource Program, Fiscal Years 1977-1986

Figure 2.31
Soil and Water Program

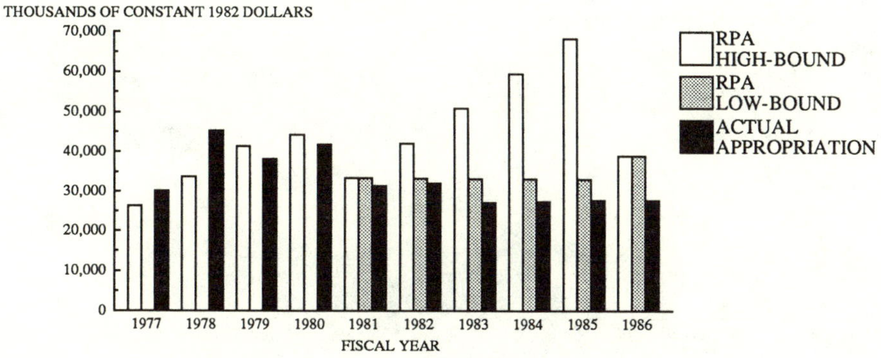

Figure 2.32
Minerals Program

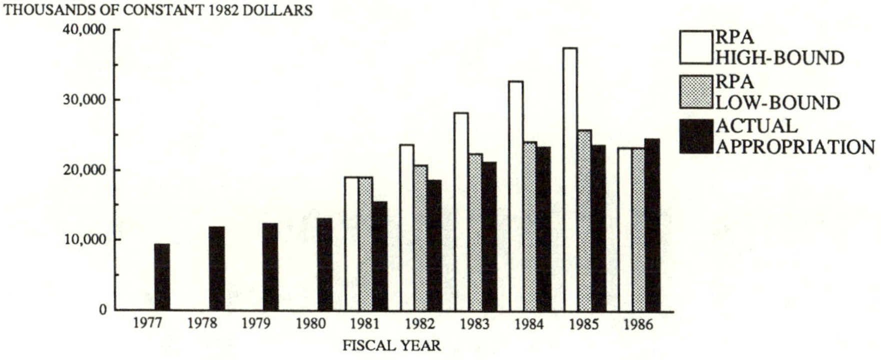

RPA Alternative Budget Targets and Actual Appropriations, by Resource Program, Fiscal Years 1977-1986

Figure 2.33
Lands Program

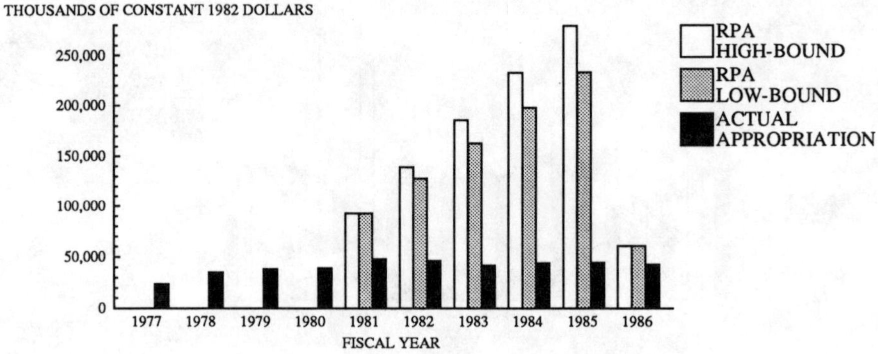

Figure 2.34
Facilities

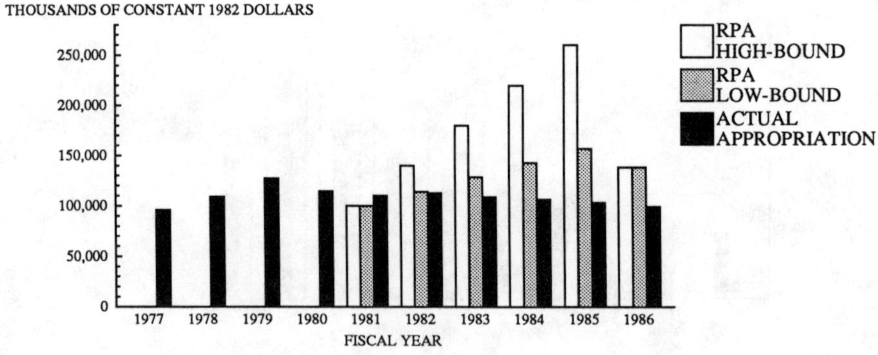

RPA Alternative Budget Targets and Actual Appropriations, by Resource Program, Fiscal Years 1977-1986

Figure 2.35
Protection

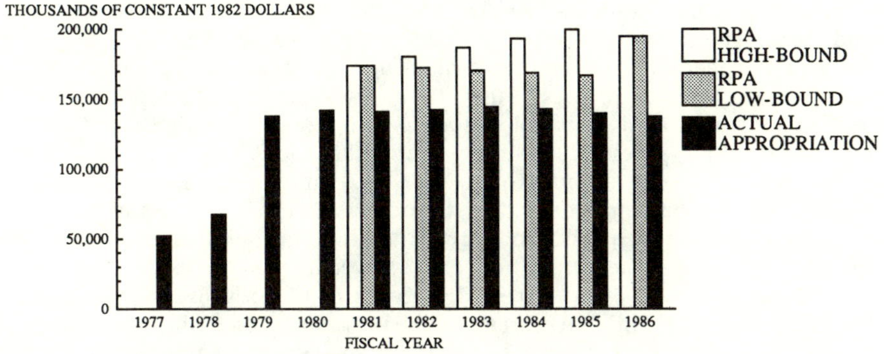

Figure 2.36
National Forest System - Total

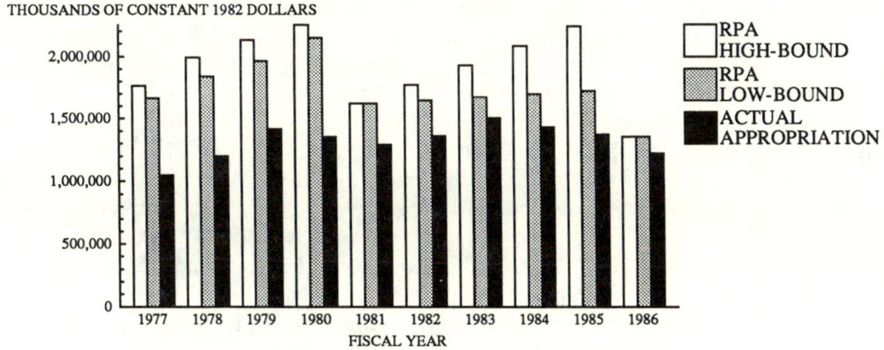

Though more piecemeal in approach than a five-year program plan, the annual budget process is well suited to OMB's formidable analytical and bureaucratic strengths.

Forest Service Program Planning: Its Influence on the Balance Among Resource Uses

A central question in this study is whether Forest Service program planning can effectively influence not only overall agency funding levels but also the *balance* of funding among resource programs. If the Forest Service is to implement *integrated* resource management on the national forests, as required by the agency's legislative mandates and represented to the public in the forest plans, then all the pieces must fit together reasonably well and in the correct proportions. The size of these "pieces," the extent of the management activities carried out within individual resource programs, is largely a function of the funding received to carry out those activities. An important measure of the degree to which such an integrated resource management plan can be implemented is this congruency between the estimated budget allocation required to carry out the various resource management activities and the actual budgets received for the tasks.

Figure 2.37 shows the distribution of resource management funding indicated in the three RPA Programs to date. Figure 2.38 shows the distribution of resource management funding in actual appropriations for the same period and, for reference and comparison, back to FY 1960. The irregularities in FY 1981 and FY 1986 in Figure 2.38, particularly noticeable in the timber program, correspond to the first year of each new RPA Program, when the Program budget goals are matched with the actual proposals in the President's Budget (this may be more clearly evident in Figures 2.27 to 2.36, which show the RPA Program goals for each resource program by fiscal year). The most noticeable trend in Figure 2.38 is the steadily *declining* share of NFS funding envisioned for the timber program from FY 1977 through FY 1980 (from 76% to 67%) and again, after the "correction" in FY 1981 through FY 1985 (from 77% to 66%) to the next correction (79%). Accompanying this, a steadily *increasing* share of funding was envisioned for several of the non-commodity resource programs, most notably fish and wildlife, and recreation. Recreation funding was seen to grow from 15 percent to 20 percent by FY 1980; at the start of the 1980 Program, the goal was rolled back to 13 percent for FY 1981, growing to 18 percent before again being rolled back at the start of the 1985 Program to 14 percent for FY 1986--lower than it had been in FY 1977.

Actual appropriations for the timber program did decline from 81 percent in FY 1977 to a low of 76 percent in FY 1980 but increased again to 80 percent by FY 1985. The proportion for actual recreation funding declined in several individual years but, overall, slightly

Distribution of Funding Among Resource Programs: RPA Targets and Actual Appropriations

Figure 2.37
RPA Budget Targets, FY 1977-1986

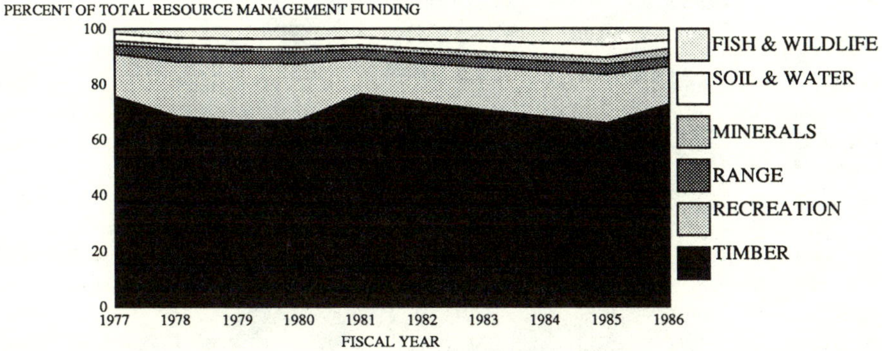

Figure 2.38
Actual Appropriations, 1960-1986

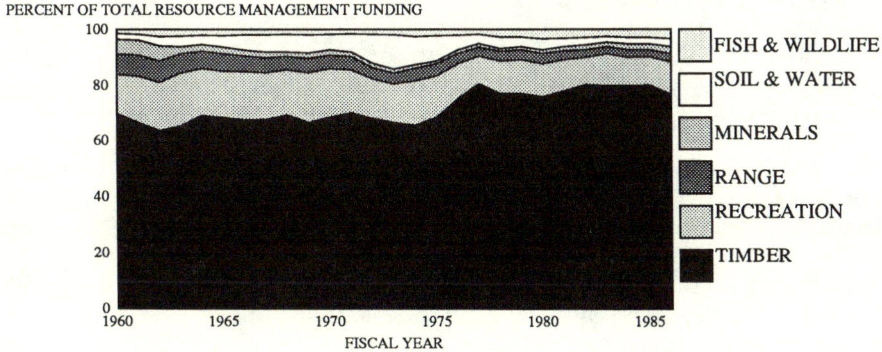

increased its share, from 10 percent in FY 1977 to 12 percent a decade later. The fish and wildlife program share increased over this same period from 1.7 percent to 3.4 percent, soil and water declined from 3.4 percent to 2.8 percent and minerals increased from 1.2 percent to 2.5 percent. Though these percentage differences seem small, they become quite significant when viewed in absolute terms. For example, the FY 1985 RPA budget goal for the soil and water program was 4.6 percent of the overall NFS budget, only 2.1 percentage points higher than its actual 2.5 percent of NFS appropriations for that year--but that translates to a difference of $22.8 million, quite a difference when one considers that the entire appropriation for this program in FY 1985 was $27.7 million.

In terms of general policy trends, the balance of funding among the various resource programs is moving away from the direction called for in Forest Service program planning, especially when viewed over the longer term. The timber program and minerals program are gradually constituting a greater and greater share of overall funding while programs such as soil and water, and recreation, which received relatively greater emphasis in Forest Service plans, are being squeezed to comprise a steadily shrinking share of the budget. It is important to note that the largest single shift away from the general direction set out in RPA planning began not with the Reagan administration, as has been widely presumed, but in FY 1977, the first year covered by the 1975 Program. In that year, appropriations for the timber program jumped 57 percent in real terms, from the prior year's appropriation of $448 million to $703 million. In the course of two years, FY 1976 and FY 1977, the timber program had gone from 68 percent of the NFS resources budget, where it had remained virtually unchanged since at least as far back as FY 1960, to over 80 percent. The Forest Service had proposed a 60 percent increase in the timber program for FY 1977 and, with solid support from the rest of the executive branch, obtained nearly all of that in increased appropriations. Similar increases were proposed by the Forest Service for the other resource programs as well but were not supported by OMB or USDA. Again, the Congress seemed to look to OMB for the best estimate of what constitutes a well-balanced and integrated resource management program on the national forests. Thus the question remains: Will the management of renewable natural resources on the national forests actually be guided by long-term goals developed by professional resource managers and publicly affirmed through Forest Service program planning? Or will it continue, as in the era before comprehensive planning on the national forests, to be driven by short-term annual budget objectives that originate outside the planning process and are decided upon in the absence of the broad public involvement envisioned in RPA and NFMA?

NOTES

1. Executive Office of the President, Office of Management and Budget, *Preparation and Submission of Budget Estimates,* Circular A-11, Revised June 8, 1987. Section 12.9 (a) and (d).

2. Ibid.

3. USDA Forest Service, Resources Program and Assessment staff, 1976. Consistent with the "crosswalk" developed by the RPA staff, budget line items were aggregated as follows:

TIMBER: timber sales preparation and administration, reforestation and timber stand improvement (both appropriated and allocated under the Knutson-Vandenberg Act (K-V), 16 U.S.C. 576), insect and disease control, access roads, advance logging, forest road construction (90%), brush disposal, timber salvage sales, Tongass Timber Supply Fund (86%), timber purchaser roads and purchaser elect roads.

RECREATION: recreation use management, forest trail construction, forest trail maintenance, recreation use area construction, and construction of recreation facilities.

FISH & WILDLIFE: wildlife and fish habitat management, K-V wildlife.

RANGE: range management and range betterment fund.

SOIL & WATER: soil and water management, water development construction, pollution abatement, water resource development.

MINERALS: minerals area management.

PROTECTION: forest fire protection.

FACILITIES: road construction (10%), road maintenance, construction for fire, administrative and other purposes (FA&O), maintenance of improvements, Tongass Timber Supply Fund (14%).

LANDS: land exchange and adjustment, special uses, land acquisition, geometronics, land line location.

4. Council of Economic Advisors, *Economic Report of the President, Fiscal Year 1987* (Washington, D.C.: Government Printing Office, 1988), Table B-3, p. 237.

5. In the bar charts, the bar marked "prior year base" for a given year often does not match precisely with the bar marked "final appropriation" for the previous year. This is due primarily to the conversion of all figures to constant dollars, and the subsequent drop in real-term value of a given appropriation level from one year to the next. In some cases, part of the difference is also due to some portion of the appropriation being either rescinded or deferred to the following fiscal year.

6. Up through fiscal year 1982, fire fighting costs were covered through an emergency supplemental appropriation. This appropriation was made during the fiscal year once the fire season had ended and the actual costs were known. Since that time, the Forest Service has operated a kind of revolving fund in which Congress appropriates funds sufficient to cover the known fire fighting costs of the previous fiscal year, and the Forest Service then uses these funds to cover fire fighting costs in the current fiscal year. This arrangement works reasonably well except when the incidence of fires in the current fiscal year is markedly higher than that in the previous year. In such a case, the Forest Service may still have to approach Congress for a supplemental appropriation in order to cover the current year obligations.

7. Fighting Forest Fires (FFF) is funded separately from forest fire protection activities. Due to the unpredictability of funding requirements, and the huge expenses involved with mobilization of both Forest Service and contracted fire crews, overtime pay, heavy earthmoving equipment, and a fleet of fixed wing aircraft and helicopters, Congress has in the past funded FFF largely through supplemental appropriations. After the FY 1976 change in the fiscal year end, from June 30 to September 30, Congress began simply reimbursing the Forest Service for fire fighting expenses in the following year's normal appropriation. Thus, the agency is permitted to obligate funds in anticipation that they will be fully covered (current obligations are paid out from appropriations made to cover the previous year's expenditures in a sort of revolving fund; in an extraordinarily bad fire year, such as FY 1987, this revolving fund may not be adequate and the Congress may provide emergency supplemental funding to cover current obligations.)

8. USDA Forest Service, *Report of the Forest Service, Fiscal Year 1985*, (Washington, D.C.: USDA Forest Service, 1986), p. 13.

9. Ibid., pp. 31-32.

10. Ibid.

11. Several different approaches to a regression analysis of the data were considered, but all were rejected due to limits on the applicability of the technique to this type of information and to limitations of the data itself. The first approach was to use an

ordinary least squares multiple regression approach to determine which of the independent variables, including "prior year base," had the greatest influence on the coefficient of determination (R^2) and the sum of the squared residuals (e^2). Time-series data is inherently subject to serial correlation so, in a second approach, the data was transformed to a "first difference" format, that is, the annual change in final appropriations, $y_t - y_{t-1}$, was assumed to be some function of the annual change for each independent variable, $x_{it} - x_{it-1}$. Because the independent variables--the actions of each of the various participants in the process, are clearly not independent of one another--the multicollinearity becomes a problem. In this case, the problem could not be adequately corrected for by expanding the data or dropping one or more of the highly collinear variables. When all this was considered, along with the fact that the most important data was limited to only a ten-year time series, it was decided that correlation coefficients (r) and a plot of the correlation with appropriations for each independent variable would be the most valid way of presenting the available data.

12. See: Wildavsky, A., *The Politics of the Budgetary Process* (Boston: Little, Brown and Co., 1964).

13. Leman, Christopher K., *Resource Assessment and Program Development: An Evaluation of Forest Service Experience Under the Resources Planning Act, with Lessons for Other Natural Resources Agencies* (unpublished paper) (Washington, D.C.: U.S. Department of the Interior, Office of Policy Analysis, August 1980).

14. For comparison purposes, federal revenue sharing programs such as Payments in Lieu of Taxes (PILT) and the 25% of Forest Service receipts paid to states and counties are not included in these figures. While the 25% payments are a direct function of the level of activity in Forest Service programs (primarily commodity programs such as timber, minerals and range), the PILT payments made by BLM are based on a flat rate per acre and are relatively constant. The data presented here reflects only the funding applied directly to renewable resource management on federal lands.

15. Leman, *Resource Assessment and Program Development.*

3

Budget Development on Five National Forests

In order to understand what influence these changes between planned and actual budgets have had on the actual "on-the-ground" management of national forest resources, a series of field studies was conducted on several national forests. The field studies serve two purposes. First, through a detailed documentation of the budget development and execution process during a single fiscal year, it is possible to see just what changes were made to the budget, and by whom, at every decision point from the development of the original budget proposals at the national forest level to the final allocation of funds to each national forest nearly two years later. Second, because such numerical data often tells only part of the story, a detailed examination of the actual management activities on each case study forest during that fiscal year reveals the degree to which the changes between the planned and actual budget materially limited or altered planned management activities. By gaining a better understanding of the functional relationship between budgets and management activities at the field level, it may be possible to better anticipate the impacts of future budget modifications on the implementation of the national forest plans prepared under NFMA.

The Five National Forest Case Studies: Methods

Five national forests were chosen for the case studies. They were chosen to be reasonably representative of the National Forest System, not in the statistical sense of representativeness, but to examine the actions of a diversity of administrative and management personnel in several geographic regions of the country dealing with diverse sets of natural resource opportunities and demands. Administrative records were used to document both the budget development and budget execution processes for each national forest though the availability of such records varied widely between forests. Personal interviews with management and administrative personnel were used to provide additional factual information as well as a depth and perspective that could not be represented in official administrative records.

Forest 1

Vermont's Green Mountain National Forest, Forest 1, is in the Eastern Region (Region 9) of the National Forest System, a geographic region covering the entire northeast quarter of the United States, from Minnesota to Maine to West Virginia (see Figure 1.1). Unlike the western national forests which were carved from the public domain, the eastern national forests were purchased by the federal government in individual tracts, largely during the 1920s and 1930s. At the time of purchase, many of these lands had been cut over, burned over, heavily eroded and abandoned--"lands nobody wanted."[1] Management of these lands was, for several decades, primarily custodial, protecting them from fire and further abuse while the land recovered through natural processes of forest growth and regeneration. In recent decades, these lands have been increasingly relied upon to contribute timber and other products as well as wildlife habitat and watershed protection. Again unlike the western forests, which occur in contiguous blocks of sometimes millions of acres, the eastern national forests occupy only a small proportion of the overall forest land base of the region, often in scattered parcels. These relatively small areas are within a day's drive of nearly two-thirds of the nation's population, however, and they have become increasingly important as recreation and scenic resources. Forest 1 itself is only several hours' drive from the largest population centers in this relatively heavily populated region. Over the past two decades, recreational use has increased dramatically and the forest is host to many winter sports enthusiasts as well as summer users. At the same time, its "third forest," following two episodes of exploitation since European settlement, has matured to diverse and valuable stands of sawtimber-size northern hardwoods and pines.

On all the national forests, the rising demand for all uses, some of them in direct conflict with one another, have made the management task of Forest Service officials more challenging and difficult. However, because of the relative lack of dependent commodity industries, or perhaps because of commonly shared social values in this relatively unspoiled corner of a highly developed region, the NFMA management plan for Forest 1 was worked out with a remarkable lack of acrimony or contention among the interest groups, and has become a model for other national forests. The management plan for this forest has been completed and implementation has begun.

Forest 2

The Jefferson National Forest in southwestern Virginia, Forest 2, is in the Southern Region (Region 8) of the National Forest System, which comprises the entire southeast quarter of the United States, from Virginia to Texas. The Southern Region largely shares the history of the Eastern Region in being comprised of lands once exploited and abandoned, acquired by the Forest Service and nurtured to their current state of health and productivity. Forest 2, located along the crest of the central Appalachian highlands, epitomizes this history. The increasing productiveness of this forest has been of benefit primarily to local users for much of its history. Mineral extraction continues to be an important factor in its management and it is an important source of raw material for the local forest products industry, from papermakers to furniture manufacturers. Recreation use in the past has been primarily from local hunters and fishermen, but the increasing size and mobility of populations east of the Appalachians and in the Tidewater area has made Forest 2 an important focus for wildland recreation.

The history of Forest 2 departs significantly from that of Forest 1 with regards to the course of forest planning. Established patterns of local use, primarily for commodities such as timber and minerals, have come increasingly into conflict with newer patterns of non-commodity resource use. Attempts to increase timber harvesting as the forest continues to mature have met with strong resistance from users who see the same mature and relatively natural forest as a unique resource in the East that should be preserved. The Forest Service is in the process of crafting a careful compromise in hopes of satisfying each interest reasonably well. The translation of this compromise into management activities is sure to be monitored very closely by all parties as the forest management plan is implemented.

Forest 3

Forest 3 is the Flathead National Forest in northwestern Montana. With over 2.6 million acres, Forest 3 is the largest national forest in the Northern Region, which occupies most of Montana and northern Idaho. It has traditionally been one of the foremost timber-producing forests in the region. It is the largest source of raw material for a dozen or more local forest products manufacturers in a region whose economy is strongly dependent upon the forest products industry. It is also on the border of a major wilderness-oriented national park, and itself contains many thousands of acres of rugged and pristine backcountry. With several neighboring forests, it shares jurisdiction over one of the most highly prized congressionally-designated wilderness areas in the country, and has the bulk of the responsibility for the protection and use of the area. Both inside and outside this area, Forest 2 contains some of the most important habitat in the continental United States for elk, moose, bighorn sheep and endangered species such as the grizzly bear and Dolly Varden trout. Recreation users come from all over the nation and from other countries to enjoy unexcelled opportunities for big game hunting, trout fishing, hiking, and camping, and this in itself has created an industry of substantial importance to the local and state economies. Although Forest 3 is in one of the most sparsely populated areas of the United States, local recreation use is an important factor as well. A large proportion of the local population uses the area for recreation, especially for big game hunting and fishing.

Because of its national constituency, the use conflicts on Forest 3 are perhaps greater than might be expected for so vast a resource base. The primary use conflicts are almost a mirror of the internal competition for budget resources between resource programs. In recognition of the uniquely valuable wildlife resources on Forest 3, there is a strong public push for greater emphasis on wildlife and fish habitat protection and enhancement. Similarly, there is a strong push for more attention to recreation management, especially the management of wildlands and designated wilderness to protect opportunities for solitude in an unspoiled natural environment, and to protect the wilderness itself from the ravages of overuse and consequent environmental degradation. Corollary to the above is a reduced emphasis on timber production, which has in turn been resisted by the local industry and by local and state officials concerned with maintaining a strong state economy.

Forest 4

Forest 4, the Carson National Forest, is in a relatively remote area of the Southwestern Region (Region 3), which encompasses Arizona and New Mexico. This area has a unique history among the western national forests in that, rather than being carved from the public domain, it was largely assembled from purchases of former land grants to Spanish colonists in what, prior to the Mexican-American War in 1848, was a northern province of Mexico. That these purchases were in most cases made indirectly from land speculators has been a major source of difficulty for the Forest Service, and is at the root of some of the most intractable issues in Forest 4's forest management planning.[2]

The northern Sangre de Cristo mountains are primarily scrub lands of sagebrush, juniper and pinyon pine with stands of spruce, Douglas-fir, ponderosa pine and bristlecone pine appearing at higher elevations. The timber industry is thus small and directed almost entirely to local markets but, because most of the merchantable timber is on the mountain crests managed by the Forest Service, it is almost entirely dependent on national forest timber. Though small, the industry is also important because it is one of the few manufacturing industries in an economy otherwise characterized by subsistence farming and grazing in small mountain villages. The wealth of population centers such as Santa Fe or Taos is based heavily on winter sports, recreation and tourism, and remains strongly localized. Although some controversy has arisen over Forest Service proposals to increase timber production, the conflict is not so much with competing uses as over concerns for soil and water quality and for the protection of a vast network of prehistoric Native American villages and ceremonial sites, many of which are as yet undisturbed. The Forest Service has been held responsible for timber harvest operations in the past that have inadvertently destroyed many of these sites and their cultural artifacts with them. The Forest Service has sought additional funding to permit the identification and cataloging of the most important cultural sites; this is one issue being dealt with in the context of forest planning.

The most serious resource issue is the level of subsistence grazing that the Forest Service will permit, which also can be thought of in large part as a budget problem. Tightly restricted levels of permitted grazing, based on the Forest Service's assessment that the range resource was in too poor a condition to support higher levels, contributed to an armed uprising in the 1960s that left two people dead and several Forest Service administrative buildings in charred ruins. The direct and immediate congressional response was to provide a massive increase in funding to create new pasture and improve existing grazing lands. The investments made, the grazing capacity increased and the issue died away--and so did the funding. Twenty years later the results of these range improvements have nearly disappeared and grazing restrictions are again a major political issue for the Forest Service managers on Forest 4. How or whether the issue can be resolved may be, to a large extent, a function of what funds are made available to implement the forest plan.

Forest 5

At nearly 2.9 million acres, Forest 5--Washington's Mount Baker-Snoqualmie National Forest--is the largest national forest in the continental United States.[3] It is also one of the most diverse in the Pacific Northwest Region (Region 6), a region known for its vast rain forests of Douglas-fir, hemlock, spruce and cedar. In addition to being a major timber producer, Forest 5 is also extremely important as the watershed for several population centers, including the largest metropolitan center in the Pacific Northwest. Being close to such a population concentration places considerable recreation pressure on Forest 5, and it is used extensively for hunting, fishing, mountaineering, hiking and a variety of winter sports, both dispersed and at developed sites. Forest 5 borders on two large wilderness-oriented national parks and has extensive areas of designated wilderness itself. In addition to providing high quality water for agricultural, industrial and domestic use, its rivers and streams provide spawning habitat for several species of salmon that are of major economic importance in the region. Its expanses of low elevation old-growth forest provide important habitat for a number of threatened and endangered plant and animal species.

The high value of these old-growth forests, both to the wood products manufacturer and the wilderness enthusiast, is the focus of most of the resource conflicts on Forest 5. One major facet of the conflict, the rate at which the remaining nonwilderness roadless lands will be entered for timber harvesting, is a direct function of the funding for timber activities and the timber output targets handed down through the budget and appropriations processes. Another major facet that will be influenced by timber harvest levels is the extent to which the Forest Service will be able to protect religious sites of the fifteen Native American tribes living within the boundaries of Forest 5. The forest plan, still in draft form at this writing, will limit timber harvesting to a level that all parties generally agree does not pose a threat to these nontimber resource uses. Whether this balance can exist in annual program funding as well as in the management plan will determine how well this fragile compromise can be upheld.

Budget Development for Fiscal Year 1986

To gain a better understanding of the extent, nature and source of deviations between planned and actual budgets, the budget development and execution processes were carefully reconstructed for a single fiscal year. The objective was to examine the detailed budget proposal prepared by each of the five national forests and to document the changes made to that proposal as it traveled up through the administrative hierarchy of the agency itself, and then to the Office of the Secretary of Agriculture, to OMB and finally to the Congress. The budget development information used and maintained by the Forest Service does not permit the direct tracking of budget proposals for individual national forests. As the budget proposals of all the various national forests in a given administrative region are aggregated to form the budget proposals submitted by each of the regional offices to the Washington office for further aggregation, it is impossible to distinguish the proposal of an individual national forest in a regional office proposal, or the proposal of a particular regional office in the proposal from the Washington office, the agency request. Reconstruction of the changes made at each administrative level during budget development was therefore accomplished through the examination of records of negotiations between forest supervisors and regional foresters and between the regional foresters and the chief.

Documentation for the changes made to the agency request by the Office of the Secretary and by OMB was available in detail, once the difficulty with obtaining disclosure of such information was overcome. The data on numerical changes in the budget proposal were supplemented with information obtained from internal memoranda and working documents indicating the nature and outcome of negotiations between the Forest Service and officials from the secretary's office and OMB. The route followed by the budget proposal, even while still within the executive branch, is a circuitous one, however, and a chronology is provided as the best way of indicating the sequence of decisions in the various feedback loops leading to the inclusion of the budget for the Forest Service in the President's Budget. Changes proposed by the various House and Senate appropriations subcommittees and committees are, of course, a matter of public record. Here again, however, the route followed is circuitous and a chronology is provided to show the sequence of decisions by the various participants in the budget process. What the chronology also provides is a sense of the variety of events, only a few of which relate directly to questions of national forest policy, that have significant impacts on both the substance and progress of Congress' appropriation of funds to manage the national forests.

The process of allocating the funds down through the administrative structure of the Forest Service, was documented through records of negotiations between the line officers at each level, and the detail for each individual national forest being studied did not emerge until the final allocation from the regional office. At this stage, it was possible to draw a direct comparison between the set of resource management activities (and associated funding requirements) thought necessary by field managers to implement the existing resource management plan and the set of activities actually funded through appropriations.

To have stopped at this point would probably have provided, in itself, some interesting and useful information on the various twists and turns of the budget process and how actual budgets come back to the field with only a vague resemblance to those originally proposed. In many public agencies there is little budget flexibility remaining at this point, and the application of funds to those activities for which they have been appropriated proceeds fairly mechanically. For the field managers of the national forests, that is the forest supervisors and district rangers, events and circumstances that will significantly influence the deployment of resources have yet to unfold when the final budget allocation is made at the start of the fiscal year. The field season has yet to begin on many national forests--late snows may significantly delay the fielding of seasonal crews, reducing personnel costs but leaving needed work undone; fairer weather may permit greater accomplishments than anticipated, but with higher-than-allocated personnel costs; a bad forest fire season may require that most of the personnel be sent on detail assignment to another national forest, leaving the manager with both unfinished work and considerable unspent funds. Thus, to gain a truly useful understanding of how budgets are actually executed at the field level, it is necessary to go beyond the final budget allocation and follow the events of the entire fiscal year through to the end-of-year accounting.

For each case study, then, the resource management activity levels and dollars in the original budget proposal from the national forest are compared with (1) the dollars and resource output targets in the final fund allocation, and (2) the actual resource activity accomplishments and reported expenditures, program by program, in the end-of-year financial statement. Thus it is possible to look at which proposed resource management activities were funded as planned and which were not, then to examine the reported actual resource management accomplishments to assess the impact that the deviation between planned and actual funding had on the implementation of the existing management plan. In such circumstances, however the numbers can tell only part of the story. In many cases the most important factors underlying the particular policy outcome will not become apparent in a straightforward examination of strictly quantitative data. For this reason, the study of the process data was followed up with a series of on-site interviews with Forest Service officials who had played some role in either the development or execution of the budget for the national forest in question. Officials interviewed included line officers, budget staff personnel and resource program staff personnel at every administrative level of the organization.

The results of the case studies on five national forests will be discussed in three separate sections. The first section will describe the budget development process within the Forest Service generally and as carried out on the five case study forests for fiscal year 1986. The second section will begin with the Forest Service agency request as submitted by the Washington office and explore the changes made by the secretary of agriculture and the Office of Management and Budget. This section will also chronicle the important and relevant events surrounding the Forest Service budget as it was debated by the appropriations committees and subcommittees in the House and Senate before becoming law. A third section will examine the processes by which the appropriated funds were allocated down through the organizational structure of the Forest Service and distributed to the individual national forests. This section will return the focus to the five case study forests and provide a detailed examination of how each field manager was, or was not, able to carry out planned resource management activities given the differences between the proposed and actual budget for that fiscal year.

The Forest Service Budget Development Process

Budget development within the Forest Service is one part of what is conceived in the Forest Service management model (see Figure 1.4) as a continuous chain of events that integrates resource management planning at the local and national levels with budget development, allocation and the evaluation of both expenditures and accomplishments. The process by which resource management plans are developed at the local level (NFMA forest plans) and the national level (the RPA Program) have been described above. What will be explored here is the process by which the goals and priorities emerging from the program plans are translated to program direction and then to the budget proposals prepared by field personnel.

The FY 1986 budget process began in mid-1983, when the Program Development and Budget staff in the Washington office began preparation of a statement of national direction and a detailed set of budget instructions to be used by field offices in developing their FY 1986 budget proposals (see Figure 3.1). The general emphasis of the statement of national direction purposely reflects the rhetoric of the current presidential administration. The budget directions for FY 1986 read in part:

> Increase revenues by applying good land management practices while minimizing costs and satisfying the objectives of the President's Economic Recovery Program. Priority is placed on programs which will help improve the Nation's economic condition while maintaining the long-term productivity. Continue emphasis on streamlining procedures, reducing administrative burdens and improving work processes in order to increase efficiency.[4]

For FY 1986, the Forest Service continued to experiment with how best to incorporate the resource management goals and the accompanying budget targets from RPA planning into the budget process. When the budget for FY 1979 was being prepared in mid-1976, the chief issued instructions that completely revised the structure of the budget proposals submitted by the field in order to make use of the program budget targets established in the 1975 RPA Program.[5] Until that time, Forest Service budget proposals were largely incremental in nature, and were based upon annual assessments of program needs rather than on any longer-term plan or set of objectives. The budget proposal structure instituted for FY 1979 placed the RPA budget targets at the center of the agency's budget development process. Two alternative budget proposals were also developed, one slightly below the RPA level and one slightly above, to indicate how alternative levels of overall funding could be most effectively distributed among the different resource programs. In order to ensure that the field's budget proposals summed to the national level resource output goals established in the RPA Program, the FY 1979 budget structure required that quantitative output targets for each resource be developed for each of the three budget alternatives. The budget proposal that was based on funding targets higher than those in the RPA Program also showed higher resource output targets, and vice versa. Having thus departed from the incremental approach of simply requesting a slightincrease over the previous year's appropriation, this new structure resulted in the FY 1979 agency request calling for dramatic increases in funding for nearly every resource program. As was noted in the discussion of the time-series data above, this approach was apparently

Figure 3.1
The Budget Development Process for the Forest Service

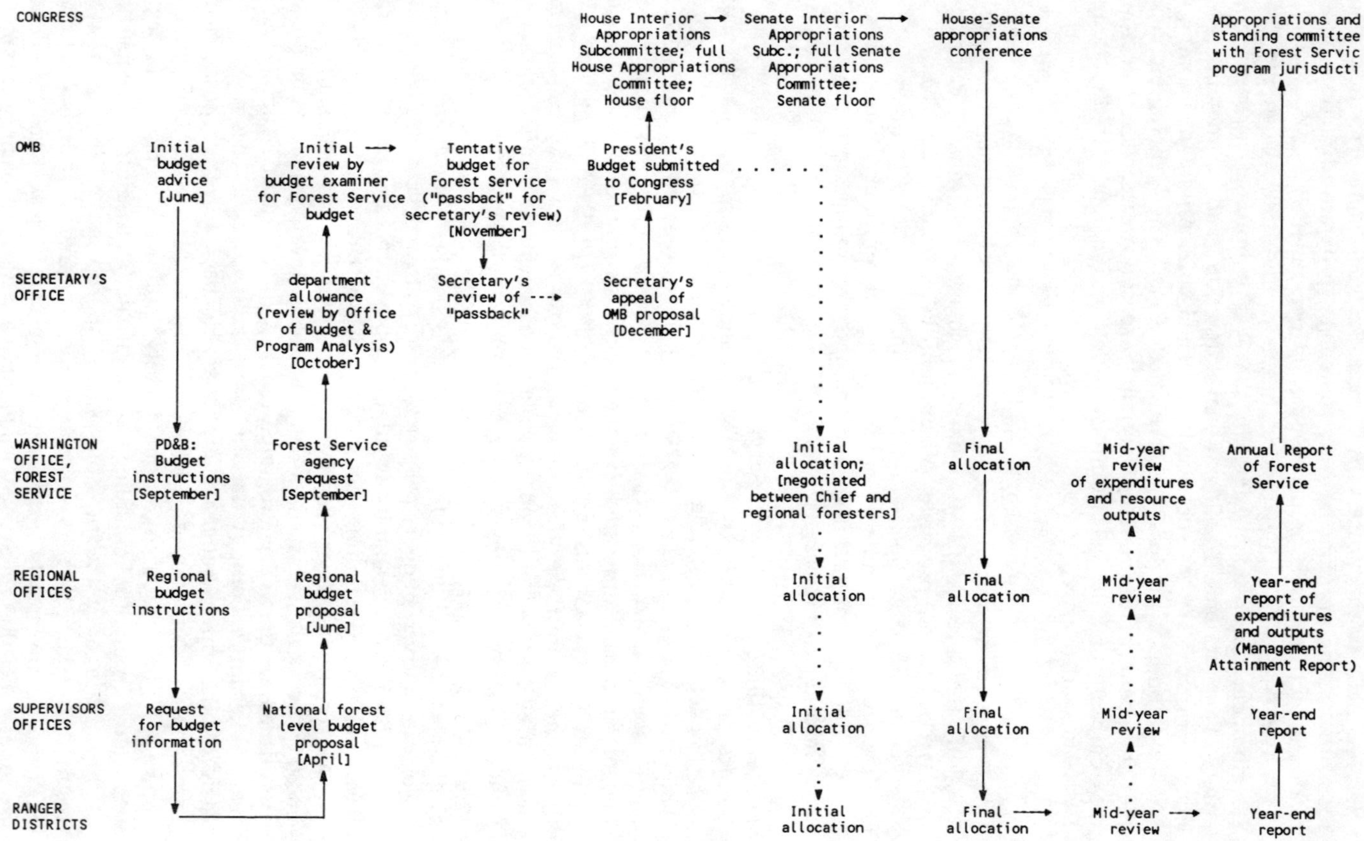

very successful in boosting appropriations for the Forest Service, and boldly higher agency requests, tied to the RPA budget targets, were used for several years thereafter. This strategy began to falter in FY 1982, however, and as it became increasingly clear that a period of overall budget contraction lay ahead, a more suitable strategy was sought.

The budget development structure adopted for FY 1986 was in effect a retreat back to incremental budgeting. The starting point for FY 1986 budget proposals was defined as "the most recent national mix of programs developed by the Washington Office and the Administration," in other words, the FY 1984 budget for the Forest Service that had recently been approved by OMB. Four other alternative proposals were to be developed--two with lower overall funding levels and two higher. The *highest* of the five alternatives generally corresponded to the budget targets of the 1980 RPA Program. A significant shift in philosophy had taken place; the RPA budget targets had gone from a central position in Forest Service budget development to defining the upper boundary in the realm of Forest Service funding possibilities. Less than a decade into its life, RPA's role and effectiveness in guiding the policies and management of the national forests had been fundamentally recast.

Beyond the general statement of national direction, specific directions for budget development within the National Forest System also echoed the themes of the current administration:

> Develop the most cost-effective budget proposals by coordinating all uses giving special emphasis to the revenue producing programs such as timber and minerals. . . . Within this guidance, develop five National proposals for timber preparation and sale: 10.0 billion board feet; 11.6 billion board feet; 12.1 billion board feet; 13.1 billion board feet and 13.6 billion board feet.[6]

Four of the five alternative proposals for the national forest timber sales level were higher than the current (1983) level of 11.2 billion board feet.[7] No corresponding national output directives were provided for minerals or for the various non-commodity resource programs on the national forests. These marching orders, along with voluminous and detailed instructions on computer coding, display layouts and interim report deadlines, were communicated directly to the regional offices, each of which prepared their own supplements for communication to the national forests.

At this point, a heretofore unified budget process differentiated into nine separate and distinct budget development processes, corresponding to the nine different administrative regions. The Forest Service is a highly decentralized organization, a characteristic that has been held by many as one of its greatest sources of strength and effectiveness, and each regional forester (the line officer in each administrative region) is given an extraordinary degree of autonomy and decision-making authority.[8] The extent to which the policies and procedures in each region differ from one another--at one time more a function of the differing landscapes, resources and socioeconomic factors among these large geographic sections of the nation--is now a function of the specialized organizational histories of each region and the individual management style and priorities of each regional forester as well. The case studies were conducted for five national forests in five different administrative regions, and the processes used to develop the FY 1986 budget proposals varied widely among the national forests studied. These differences will be discussed as the budget development process on each case study forest is examined in turn. The discussion of the process used on the first case study forest will serve as a point of departure; it will present factors common to all the case studies as well as details specific to the first. The discussions of the four remaining case studies will concentrate primarily on those aspects in which they are different from the first.

FY 1986 Budget Development on the Five Case Study National Forests

Forest 1

The FY 1986 budget development process began for Forest 1 when the Region 9 office in Milwaukee received the detailed budget instructions from the Washington office in mid-September 1984. The Region 9 budget staff provided each national forest supervisor's office (SO) in the region with a detailed report of that forest's actual fund allocation for the benchmark fiscal year 1984, arranged by "decision variable." The decision variables correspond directly to what will hereinafter be referred to as the resource programs. For example, the recreation resource program includes all budget line items pertaining to recreation activities and facilities: recreation use management, trail maintenance, trail construction, and recreation construction. The data for FY 1984 actual funding was reviewed at the supervisor's office for accuracy and completeness and submitted back to the Region 9 budget staff. The corrected FY 1984 data then became the basis for Region 9's FY 1986 "starting point" budget alternative.

In September 1984, Forest 1 was well along in its forest planning process and was already in a position to base its FY 1986 proposal at least roughly on the management activities and emphasis envisioned in the plan. Washington office instruction to the regions, however, was to build incrementally on the President's Budget for FY 1984, and this direction was faithfully transmitted from the regional office down to each forest supervisor. National direction also required the display of "a range of costs and outputs" for each resource program, noting that "the process in past years did not always provide sufficient flexibility to address the Department or Executive concerns or desired shifts in emphasis."[9] Data from Forest Service program planning was used to develop the four alternatives to this "starting point," two each above and below. The portion of the RPA national goals assigned to Region 9 were further disaggregated to the national forests by using actual data to determine the percentage each national forest typically contributed to the regional total, then disaggregating the regional RPA goal for each resource program on that basis. The regional office developed all five funding and accomplishment increments for each national forest, by resource program, then sent this information to each forest supervisor for "fine-tuning." An accompanying memorandum noted, "we want to assure you that we are not trying to build your programs for you."[10]

The regional total for each resource program was then translated to the format used by the Washington office's Program Development and Budget staff. These "MIH codes," called this because they are prescribed in the Management Information Handbook compiled by the Forest Service accounting department, disaggregate the fourteen decision variables into 26 or more codes, corresponding roughly to the budget line item format in which the overall agency request must be presented to OMB and Congress. The translation is never precise, nor is it exactly the same from one year to the next. During the fund allocation process following the signing of the appropriations bill, this budget line item format must be translated *back* to "management codes" that integrate segments of several different program areas into specific multiple-use projects at the national forest and ranger district levels. There are many opportunities in this series of translations and retranslations for losing the details and nuances of carefully crafted proposals based on the field's integrated approach to resource management. The continuing inability of the budget process to facilitate decisions using the same conceptual basis that the resource managers employ in the field is a surprisingly effective barrier to improving the integration of Forest Service program planning and the budget development process.

After the initial translation of the entire budget data base to MIH codes, the forest supervisors are then given a chance to review the proposals and suggest changes. The focus is explicitly *not* on questioning the validity of the translation. Directions sent from the Region 9 regional forester to forest supervisors with the review draft of the budget noted that "we recognize this process will not exactly fit the situation for each Forest, but we do expect this process to give us ballpark figures that each Forest can review and define."[11] There are other limits on the changes that can be suggested by the forest supervisors. The "starting point" alternative based on the FY 1984 actual allocations was required to be the same as the actual FY 1984 funding and target levels assigned to each national forest, that is, the forest supervisors could not rearrange funding among resource programs or specify different resource output targets to reflect what they thought their particular forest may need. This constraint was not imposed by the regional office, but by the Washington office; noted one regional official, "we specifically asked the Washington office to allow us to increase some functional dollars that were extremely low for our region, but they said no."

An additional constraint was that capital investment projects--construction of roads, trails and other facilities--could not be included in the initial budget proposals reviewed by the forest supervisors. All capital investment projects were to be evaluated separately and, in a departure from the doctrine of decentralized decision making, all authority for review and approval of capital investment projects was withheld by the regional office, not passed down to the forest supervisors. The basic information on capital investment needs had to be generated in the field, of course, but even there this part of the process was handled separately and with special care. First, the district rangers were asked by the forest supervisor to submit proposed capital investment projects to the supervisor's office for review. The forest supervisor then crafted his or her own strategy for obtaining funding along the lines of his or her own sense of priorities. These were submitted to the regional forester as a series of alternatives, along with a detailed narrative describing the consequences associated with not following through with the recommended alternative.

For FY 1986, there were specific upper limits on funding for each resource program and lower limits on resource output targets (primarily on timber sale levels) specified by the Washington office and transmitted to the national forests through the regional office. The regional office was keenly aware of the competition for funding that would take place between all the regional offices later in the process, and took several steps to improve its competitive position. Each SO in the region was asked to submit data on past spending and accomplishments so that the costs of production could be compared for the different national forests. For this year at least, the regional office ran a series of regressions to identify those national forests showing higher than average unit costs of production for the various resources. It was made clear to the forest supervisors that they would be competing with one another and that those units showing "high unit costs will be at a disadvantage."[12] It is unclear to what extent this analysis was used for making reductions on individual forests since the direction also noted that the decisions would also be based on "management direction, market conditions and political considerations." Unit costs based on actual performance among the Region 9 national forests were then applied to the assigned outputs to obtain an overall funding level for each resource program. If the resulting totals exceeded the assigned budget constraints for the region, the unit cost estimates were scaled back and different unit costs were assumed for each national forest. In addition, noting that "a few Region-wide adjustments were made to make Region 9 competitive with other regions," the regional forester made an across-the-board reduction in the funding sought for timber sales and road construction while leaving the accomplishment targets unchanged. Thus adjusted, Region 9's five budget alternatives, along with the necessary data and narratives to explain the alternatives and regional forester's recommendations, were submitted to the Washington office in May 1984.

Forest 2

The FY 1986 budget development process on Forest 2 was characterized by an unusually low regard for the utility of this portion of the budget process to field managers. More than on any other national forest studied, the managers expressed a common view that, because of the gross differences between proposed and actual funding distributions that had occurred in the past, the preparation of a field budget proposal was little more than a perfunctory exercise and had little relevance to the budget allocations that would subsequently be received. The overall budget process was seen as perhaps the most critical factor in determining which management activities would take place on the national forest, but the managers felt generally that the events that really mattered took place during the allocation stage of the process, not the development stage. An examination of the particular history of Forest 2, and of the Southern Region (Region 8) generally, suggests that this attitude may be somewhat justified. It also suggests that the Southern Region may have its own particular set of organizational obstacles to be overcome before the budget process can be effectively integrated with the program planning processes.

The current level of disregard for the process of developing field budget proposals is, to some degree, a reaction to having unsuccessfully experimented with a process that placed a great deal of emphasis on this portion of the budget process and required enormous effort on the part of field personnel. In the early 1980s, the Forest Service developed a computerized system for producing budget proposals that were firmly grounded in detailed estimates of costs and resource activities, on a project-by-project basis. Known as ADVENT, this system required field managers, including ranger district staff, to anticipate two years in advance all of the activities that would involve expenditures on personnel or materiel. Detailed work plans had to be prepared for nearly every project that would be undertaken, and all activities and investments "costed out." The input requirements for this program were extremely intensive and consumed the efforts of a number of personnel on every ranger district and national forest for weeks or months. The experience on Forest 2, which was shared to some extent by nearly every unit in the National Forest System, was that entire projects and even sets of projects were summarily swept away by gross differences between proposed and actual funding. A particular problem was the shifting of the balance of funding among resource programs such that the balance in the actual funding was grossly mismatched to the integrated projects that had been developed. Not only was much of the time-consuming work tossed into the rubbish, but in addition, new project plans had to be worked up to reflect the actual available funding and the way it was proportioned among the different resource programs. Given this, it was not worthwhile to invest time in detailed project planning until field managers had a reasonably good sense of the budget resources that would actually become available to them. A clamor arose among the field offices of the Forest Service that they be relieved of this onerous and useless burden. The ADVENT system was soon officially abandoned.

The reaction on Forest 2 and in the regional office to this signal was to take additional steps to reduce the field's responsibilities in budget development. At this point, late 1983, a tremendous amount of staff time was being dedicated to analysis and formulation of the first forest plan on Forest 2 and other national forests in the Southern Region. It was suggested that, given the Washington office instructions to develop the FY 1986 budget proposal as a function of FY 1984 actual funding, this could be done as well by the regional office budget staff as by the field offices. The forest supervisors asked that the Region 8 staff put together a budget proposal that could simply be checked for accuracy by the SO staffs and then sent along to Washington. This is what was done. Additional data was requested of the SO by the regional budget staff in order to construct the incremental alternatives below the FY 1984-based "starting point" and there was input from the forest supervisor on the contents of these alternatives. For the most part, however, the FY 1986 budget proposal for Forest 2 was created in the regional office.

Forest 3

National forests in the Northern Region (Region 1) had difficulties with the ADVENT system similar to those experienced in Region 8. The similarity was in the nature of the problems, however, not the degree. Actual funding had varied somewhat from proposed funding in the past, but the balance of funding among resource programs had generally reflected the balance in the proposed budgets to a greater extent than in Region 8.

The post-ADVENT reaction on Forest 3 in the Northern Region was correspondingly less extreme than on Forest 2. The district rangers and staff were largely relieved of budget development responsibilities, but the supervisors' offices were very involved in constructing their own budget proposal alternatives rather than leaving this task solely to the regional budget staffs. Also, the forest plans for many of the national forests in Region 1 were well along and the SO budget staffs were able to utilize the vast body of data compiled for planning purposes to guide them in preparing the budget alternatives. This was the first year that such information was available and used in the budget development process. However, the information was incomplete and the actual course of action to be selected in the draft plans still undecided. It is anticipated that the forest plans will play a much greater and more distinctly defined role in budget development in later years than they did in FY 1986.

On Forest 3, the draft forest plan was nearing completion in the fall of 1983, and the SO staff was reasonably confident they knew the general direction that would be taken in selecting among the planning alternatives. While there were no project plans on which to base estimated costs and resource outputs, the plan did provide guidance on the relative emphasis that would be placed on the different resource programs. The "starting point" budget alternative was developed based on the FY 1984 initial allocation, as per directions from the Washington office, but the remaining four alternatives were developed by the SO staff largely on the basis of the nascent forest plan. Once these five alternatives had been constructed, there was a convening of all the district rangers to discuss the assumptions on which each of the alternatives rested and to allow the rangers to request adjustments in the proposals. The adjustments were negotiated between the district rangers and the forest supervisors, and between the district rangers themselves, and were firmly decided upon by the end of the meeting.

The regional budget staff accumulated the budget proposals from the fifteen national forests of Region 1 and then adjusted those proposals to fit collectively within the funding and target constraints assigned to Region 1 by the Washington office. The first task of the regional budget staff was to identify those areas in which the aggregated SO proposals fell short of targets or, exceeded funding limitations. This was done on a resource program-by-program basis, but the regional forester sought to maintain the integrated approach represented in the SO proposals by consulting with the forest supervisors before making specific reductions. The forest supervisors met with the regional forester and his or her budget staff, some individually and some in groups (usually groups of neighboring national forests) to obtain a rough consensus on how the necessary reduction in a particular resource program could be spread around among the fifteen national forests while doing the least violence to the integrity of the original proposals. Here again, one factor considered was the unit costs of production on each forest: The regional forester was looking for forests showing higher than average unit costs for a particular resource program as a way of identifying opportunities to reduce spending by more than the reduction in outputs. The respect for integrated resource management often presented countervailing considerations, however. For example, some of the national forests in the dry, rolling hills of eastern Montana showed higher than average unit costs for timber production due to lower stand densities, smaller average stem diameters and lower product values; nonetheless, timber production was judged necessary for carrying out important big game habitat management activities and thus was continued despite the higher unit costs. With the major issues ironed out more or less to the satisfaction of the forest supervisors, the Region 1 budget alternatives were transmitted to the Washington office.

Forest 4

The FY 1986 budget development process on Forest 4 was, in effect, a hybrid of the processes described above for Forests 2 and 3. The budget staff at the Southwestern Region (Region 3) office developed the FY 1986 budget proposals for the twelve national forests in the region without any prior submittals at all from the supervisors' offices themselves. Again, in part due to the wake left by the ADVENT experiment and in part due to the extraordinarily demanding workload placed on the supervisors' offices by forest planning, the regional office assumed the primary responsibility for budget development, and they apparently got little argument from the field. The reason given by the regional forester was that he "wanted to maintain the sense that the forests are not standing alone, but are part of a regional team."

Once the tentative regional budget proposal was completed, all of the forest supervisors were brought in as a group to review it with the regional forester and budget staff. There exists very little documentation for this process. First, Forest 4, and presumably the other forests in the region as well, truly had no proposal of their own to submit. An exhaustive search through the budget records at the supervisor's office turned up no computer records or analyses, and no paper records except for two handwritten pages showing rough dollar estimates of the costs for the major programs. There were no estimated resource outputs specified for Forest 4 or for any other forest in the region. There was apparently no involvement whatsoever from the district rangers or their staffs. The Region 3 budget staff had, in accordance with Washington office direction, used the funding and resource output targets from the FY 1984 initial allocation to construct their central budget alternative, but there was no further effort to disaggregate these funds or targets among the twelve national forests.

With more of a sense of the direction emerging in their forest management plans than anything concrete or quantitative, each forest supervisor came to the regional office to offer suggestions and make adjustments to the tentative regional proposal. Changes seem to have been proposed strictly on the basis of individual concerns or preferences for funding priorities. The focus of the meeting was the projected display of a computer screen showing a spreadsheet with all of the budget line items for each resource program. Line items were discussed one by one in open debate around a large conference table, and when consensus was reached among all the forest supervisors and the regional forester the change was keyed in and appeared immediately on the display before them. A running total was recalculated automatically with each new entry so that the adjustments did not push the total over the regional funding allocations for each budget alternative. Since these changes were made "in real time" and without interim or periodic printouts, few records exist of the negotiations beyond the individual recollections of the attendees. The product of this session, thus negotiated and "validated" with the forest supervisors, went through additional minor adjustments at the regional office and was submitted to Washington as the regional budget request.

Forest 5

FY 1986 budget development on Forest 5 was similar in many ways to that on Forest 3. Although many of the field managers on Forest 5 had felt that the labor input requirements of the ADVENT system were a bit excessive, deviations in the balance of funding between the proposed and actual budgets had not been so great as to invalidate large numbers of the completed project plans that served as the basis for the ADVENT data base. In monetary terms, the timber program on Forest 5 is far larger than any other resource program, and a significant percentage increase or decrease in any of the nontimber programs could take place without resulting in a readily visible difference in the overall distribution of funding. Because the use of ADVENT had actually proved quite useful to managers on Forest 5, a scaled down ADVENT-like approach was still used by several district rangers and the SO budget staff to help develop annual budget proposals, even though this approach was no longer officially required, nationally or regionally.

The resources under the purview of each district ranger on Forest 5 are comparable, in value if not in geographic extent, to those overseen by many forest supervisors in other regions. Generally, the district rangers are deeply involved in the annual budget development process, and this was the case in FY 1986. Each district ranger was asked to submit a proposed program of resource management activities and expenditures, and they found it convenient to use an ADVENT-like approach to develop this. The SO budget staff reviewed these proposals, requested additional information from the districts where necessary, and constructed a forest-level budget proposal based on the priorities of the forest supervisor. There was close communication between the supervisor's office and the Pacific Northwest Region (Region 6) office even at this early stage, perhaps to ensure that the task at the regional office would involve only fine-tuning and not major rewrites of a budget representing many millions of dollars in resource values. Once a tentative Forest 5 budget proposal was complete, the SO budget staff met individually with the district rangers to iron out major differences and to be sure the data underlying the proposal was as up to date as possible before being submitted to the regional office.

The process at the regional office was largely a repeat of the process at the supervisor's office. Budget alternatives submitted by all of the national forests in Region 6 were used to compile a tentative regional budget proposal. Where cumulative totals for individual resource programs were found to be in excess of Region 6 funding constraints or to not meet resource targets, the regional forester and budget staff negotiated with the forest supervisors, in some cases individually and in other cases in small groups, to find the most efficient and practical means for increasing outputs while reducing costs. One problem arose in developing the regional proposal that was particular to FY 1986 and, for the most part, to the Pacific Northwest Region. In the fall of 1984, Congress passed legislation to allow timber purchasers to opt out of timber sale contracts which, due to the recession in the wood products markets, the companies could not fulfill. The timber sales that were turned back to the Forest Service could subsequently be offered for resale, substituting them for new timber sales that would otherwise have been offered. The volume of timber turned back for reoffer in FY 1986 was large enough to substitute for a considerable portion of the anticipated new timber sales. Because most of the sale preparation work had already been accomplished on these reoffered sales, it was assumed that the annual sales volume could be held at the same level as the previous year but at much lower cost. The costs of sale preparation are primarily personnel costs, however, and the regional forester was put in the position of having to eliminate hundreds of positions for highly-trained but underutilized foresters. Not only would this be unpleasant for those individuals let go, but the regional forester knew that the reoffering of old timber sales was no more than a two- or three-year phenomenon, after which he would again need these foresters again. A great deal of effort in the Region 6 budget proposal for FY 1986 was involved with finding some way of retaining these valuable employees within the assigned budget constraints.

FY 1986 Budget Development at the Washington Office

Much of the budget development work in the Washington office is done not by the Program Development and Budget (PD&B) staff, but by "budget coordinators" representing each of the three major branches of the Forest Service (National Forest System, Research, and State and Private Forestry), plus Programs and Legislation and Administration. During much of the budget preparation, PD&B serves a coordinating and facilitating role, allowing the executives in each major program area to negotiate their own budget priorities.

In early spring 1984, several people from the field were detailed to the Washington office to constitute the "NFS Budget Team" for the FY 1986 budget. A new budget team is established each year and is dissolved once the budget has been completed and folded into the overall agency request. The NFS budget team for FY 1986, as in most years, consisted of two former forest supervisors and three specialists from particular function areas (in this case, timber, recreation, and fish and wildlife). This team worked directly with the NFS budget coordinator, who is typically on a longer-term assignment to the Washington office. The NFS budget coordinator occupies a critical position, and is seen by many in the organization as having power and influence behind only the chief and the deputy chief for the National Forest System, to whom the NFS budget coordinator reports directly. The budget coordinator is typically a senior executive who has served as a forest supervisor on one or more national forests, and often as a close staff assistant to one of the regional foresters. The selection of an individual to serve as NFS budget coordinator is often the sign that he or she has been anointed to become, after a two- or three-year stint in the Washington office, one of the nine regional foresters. As pointed out earlier, the regional foresters are second-level line officers, below only the chief, and wield perhaps more direct program authority than any other individuals in the entire organization.

Early in the process, the NFS budget team concentrated on validating the data flowing in from the regions, checking for obvious errors, correcting those that involved such problems as misplaced decimal points or typographical errors, and checking back with the regional offices to correct other errors with less obvious solutions. The PD&B staff serves several functions at this point; the first is to actually receive and consolidate the data being submitted by the regional offices, and the second is to provide guidance and assistance to the budget team itself, several members of which may have had no prior experience whatsoever in budget preparation. Once the budget data base had been assembled with a reasonable degree of confidence that the numbers were correct, the budget team turned its attention to compiling NFS-wide budget alternatives from the five budget alternatives submitted by each of the nine regions. The budget team was instructed to assemble only three budget alternatives: a continuation of the current level of services (corresponding to the FY 1984-based "starting point"), an increased level (corresponding to the RPA-based second increment above the starting point) and a reduced level (corresponding to the second decrement below the starting point). The reduced level was geared to be just above what the regional offices felt was the minimum level of activity that would permit the Forest Service to fulfill its legislative mandates for resource management and protection.

The bulk of the budget team's efforts went into fleshing out the current level budget alternative. Target ranges, based on those in the program budget guidelines, were set for several resource outputs, primarily those related to timber sales and reforestation. Target ranges were added for other non-commodity resource outputs such as recreation (measures of recreation use capacity), fish and wildlife (number of acres of fish and wildlife habitat improvement) and soil and water (number of acres of soil and water protection and inventory). Total dollar constraints had been established for each resource program, and it was the task of the budget team to distribute these funds among the nine regions. Unless there was some overriding consideration based on a significant change in the management situation in a particular region, this distribution process was consistent with the actual fund allocation among the regions in FY 1984. A similar process was used to determine the distribution of funds among the resource programs and regions for the other two budget alternatives. The three budget alternatives then became the focus of discussions and negotiations between the deputy chief for NFS, the NFS budget coordinator and, as the process neared completion, the chief himself. Both the NFS budget team and the Washington office PD&B staff provided staff assistance, checking facts and figures with the regional offices, providing additional analysis, and projecting the results of various scenarios proposed by the decision makers. The task at this point in the process was to come up with a single recommendation for funding the management of the national forests in FY 1986, to be folded into the overall Forest Service budget request.

The development of a single budget proposal for the National Forest System consisted of two simultaneous distribution operations. The first was to determine the balance of funding among the different resource programs, and the second was to determine the distribution of funding among the nine regions. The budget team still had all of the regional data at hand, that is, five alternative levels of funding for each resource program. The total range of options available to be evaluated thus consisted of 45 possible combinations: five alternative levels of activity in each of the nine regions. This assumes, however, that the integrity at each level in each regional proposal, that is, the relationship of the resource program activities to one another, would be maintained. This was not the case. As the top executives, each with their own notion of priorities, began devoting nearly all of their attention to budget development, and as social and political factors came into explicit consideration, a great deal of mixing and matching began to take place.

The baseline proposal presented to the executives by the NFS budget team was that combination of funding by resource program and region that maximized outputs in each resource area while minimizing costs. What emerged from the baseline proposal was a ranking of region/resource program combinations, with the top-ranked combinations being the highest priority from an efficiency and effectiveness standpoint. With the subsequent inclusion of political considerations, the upper ranks were significantly rearranged and new sets of resource program/region combinations were spawned on an ad hoc basis. Forest Service intelligence gathering had identified issues of particular concern to the assistant secretary, to OMB and, in particular, to key members of Congress. Funding combinations that had been fairly far down the list, or not on the list at all, found their way to the top and were assigned a high priority.

Overall funding constraints still had to be respected, however, and much of the additional analysis by the budget team, often in direct consultation with the line officers and resource program staffs at the regional offices, was on new combinations generated by the executives. It is important to note that during the field office's work on the FY 1986 budget proposal, the FY 1984 actual appropriation was the most recent available expression of the political will of the Congress relating to Forest Service appropriations. By the time the FY 1986 budget proposal reached the Washington office in May of 1984, the President's Budget for FY 1985 was already out and the congressional appropriations committees had largely signaled their intended responses. Thus, the Washington office budget strategists were working with updated intelligence and were able to tune the agency request to achieve the best possible proposal. The simple fact that budget development for any given fiscal year begins in the field a year earlier than in the Washington office, and is thus based upon data a year older, may account for some of the differences between a simple aggregation of the regional requests and the actual agency request.

The final result of the strategy and negotiation sessions at the Washington office was a substantial mixing and matching of funding by resource program and by region. Ultimately, the funding proposal for a particular region represented the starting point for some resource programs, the first decrement for others and the first or second increment for still others. Rather than maintain each region's proposal for each of the five alternative budget levels as an integrated or indivisible whole, the five alternatives had been exploded into their constituent parts and the components reassembled according to the priorities and the budget strategies of the Washington office executives, based on their own reading of the political obstacles and opportunities ahead. Nonetheless, it is important to note that the first major modification of the Forest Service field budget proposals, and the first major departure from the field's integrated resource management approach, comes not from Congress or elsewhere in the executive branch but from within the Forest Service itself.

Modification of the Forest Service Budget Request by Organizations External to the Forest Service

Chronology of Interactions Between the Forest Service, the Office of the Secretary and the Office of Management and Budget

Once the Washington office has completed a single budget proposal for all Forest Service programs (usually by September), this "agency request" is formally submitted to the Office of the Secretary of Agriculture for review and modification (see Figure 3.1). The Forest Service request is directed to the assistant secretary for natural resources and environment, who has the direct line authority in the Office of the Secretary for Forest Service programs. Providing analysis of the proposal for the assistant secretary is the Office of Budget and Policy Analysis, a staff of career analysts and economists within the Office of the Secretary of Agriculture. The modified budget proposal for the Forest Service, known as the "Department Allowance," is submitted in October to the Office of Management and Budget in the Executive Office of the President. The proposed budget for the Forest Service is directed to the politically appointed associate director for natural resources, energy and science and is analyzed by a career budget examiner permanently assigned to review Forest Service budgets. It is not unusual for a single budget examiner to have sole responsibility for reviewing Forest Service budget proposals for several years, so an OMB budget examiner can come to know and understand the programs of the agency in considerable detail and with a historical perspective equal to that of many senior executives in the agency itself.

Upon receiving the Department Allowance for the Forest Service, the budget examiner will spend a great deal of time with budget officials at the Forest Service, checking cost estimates and examining justifications for particular programs or projects. Once the examiner is satisfied, he or she will make recommendations for further modifications, based on both the agency request, the Department Allowance and whatever additional investigation was done. These recommendations are submitted to the deputy associate director for the natural resources, energy and science section, the highest career level in OMB, who further probes the agency's justifications for its funding request. If the budget examiner cannot provide a satisfactory justification for a particular request, it is likely to be reduced or eliminated by the deputy associate director. So it is in the Forest Service's own best interest for it to provide thorough responses to the examiner's earlier questions and give the examiner the best possible information for justifying the agency request inside OMB. The tentative OMB budget for the Forest Service is then sent back to the secretary of agriculture to provide the secretary's office an opportunity to review OMB's changes and appeal any that the secretary finds unacceptable. This "passback" is directed to the deputy secretary at USDA and it is his or her responsibility to coordinate with the many assistant secretaries who oversee the various agencies within USDA and to represent them in appealing any unfavorable decisions back to OMB. Appeals are normally negotiated between the Agriculture deputy secretary and the OMB associate director with responsibility for that agency, political officials roughly equivalent in rank. When agreement is reached, the office of the secretary provides any additional budget detail that may be necessary and the information is submitted directly to OMB by computer for inclusion in the President's Budget, released in January of each year.

Following is a chronology of the events and decisions in this executive branch review process that are relevant to the Forest Service budget for FY 1986.

September 1984: The Forest Service agency request is submitted to the assistant secretary of Agriculture for Natural Resources and Environment John B. Crowell, Jr. for his review and for analysis by the Office of Budget and Policy Analysis (OBPA), a staff group within the Office of the Secretary of Agriculture. The Forest Service proposed timber sale level is 11.2 billion board feet (BBF), the same level provided in the FY 1985 budget. The rationale for this level is that there will be an anticipated 0.5 BBF of defaulted sales to reoffer due to the inability of some companies to operate high-priced sales now under contract. Thus, a total of 11.7 BBF would be offered for sale, the same as the volume expected to be harvested. Housing starts have continued their slow increase during FY 1984 from the very low levels seen in 1981 and 1982. The most recent OMB economic assumptions project 1.8 million housing starts in FY 1985 and a leveling off at 2.0 million in future years. However, prices bid for Pacific Northwest national forest timber sales have not responded to this improvement for a number of reasons

1. Industry uncertainty over short term economic conditions; notes an internal OBPA memorandum "unlike USDA, timber companies are not bound by OMB assumptions and sometimes base decisions on gloomier forecasts."[13]

2. The surge of Canadian timber into the U.S. market, taking advantage of the relative strength of the U.S. dollar relative to the Canadian currency; it is estimated that the Canadians now hold 40 percent of the U.S. market for finished lumber and plywood.

3. The fact that much of the improvement in the housing industry has been in the South, drawing from regional wood products manufacturers who rely largely on industrial and nonindustrial private lands for timber supply.

The Federal Timber Contract Payment Modification Act of 1984 will permit holders of high-priced timber contracts in the Pacific Northwest to get out of some of their contracts, allowing the timber volume in the turned back sales to be reoffered as part of the overall national forest timber sale program in FY 1986-1988.[14] Forest Service expenses associated with the reoffered volume are expected to be somewhat lower than with "new volume" because much of the road construction and other sale preparation activity has already been done. The Forest Service estimate of the volume that will be available for reoffer in FY 1986 is 0.5 BBF. There exists a wide margin of uncertainty, however, and an overestimate could significantly increase the assumed cost of meeting the overall 11.7 BBF timber sale commitment.

October 1984: The first draft of the Department Allowance is submitted to the Office of Management and Budget for review by Frederick N. Khedouri, associate director of the natural resources, energy and science section. A USDA staff paper notes that "setting the timber sales level is the single most important decision in determining Forest Service funding and staffing."[15] Two issues predominate

1. The sale of national forest timber at a net financial loss to achieve objectives such as maintaining the economic stability of timber-dependent communities and improving wildlife habitat--"concentrating the timber sale program in the Pacific Coast states and the South would produce a more efficient program from the standpoint of generating profits for the Treasury."

2. Practical limitations on the assistant secretary's priority of increasing national forest timber sales levels. Despite the high level of housing starts called for under the administration's economic assumptions, the volume of sold but uncut timber volume on the national forests is approaching an unprecedented 40 BBF--a three- to four-year supply. Further, under orders from the White House, the Forest Service has reduced personnel by 5,250 staff years since 1981; a large increase in the timber sale level will not be possible without significant increases in staffing.

In a decision memorandum prepared by the Office of Budget and Program Analysis, two alternatives to the Forest Service request are presented to the assistant secretary

1. Increase timber sales from the 11.2 BBF proposed by the Forest Service to 12.2 BBF. Timber sale costs would be increased by $68.7 million; but if Research were decreased by $10.8 million and State & Private Forestry were reduced by $19.2 million from the agency request, the net increase would be only $38.7

million. Over 70 percent of the additional funding would have to go to high-cost regions, which would produce only 43 percent of the increase in sales volume. "The more cost efficient regions are near their output capacity at the agency request level, and it is not possible to assign large increases in sales to these areas."[16]

2. Maintain the sale volume proposed by the Forest Service, but redistribute timber sales to obtain greater economic efficiency. The agency request maintains the same regional distribution of timber sales as in prior years. Redirect sales to the Southeast and Pacific Coast, operating sales in other regions at 80 percent of prior years' level. Costs would be $19 million less than in the agency proposal and value of the timber sold would be $109 million more, yielding a greater return to the Treasury. Personnel would be reduced by 200 staff-years from the agency request. The Rocky Mountain and Intermountain areas would have the greatest reductions and Colorado and Idaho would probably be the states most adversely affected; "this would present obvious problems since Senator McClure (R-Idaho) chairs both the Energy and Natural Resources Committee and the Interior Appropriations Subcommittee."[17]

Crowell's decision in the Department Allowance, as submitted to OMB, is to choose the first alternative, a harvest level of 12.2 BBF.

November 1984: The OMB "passback" to the secretary of agriculture outlines changes made to the Department Allowance for the Forest Service. Timber sales level is reduced from 12.2 BBF to 10.3 BBF, nearly one billion board feet below the agency request and two billion board feet below what the assistant secretary sought.[18] The secretary's allowance for direct funding for timber sale preparation was reduced nearly $33 million (15%) and funding for road construction for timber access was reduced from $246.5 million to $146.4 million (41%). Minor reductions in nontimber resource programs were largely to separate out the costs of a new service-wide computer system that in the agency request had been apportioned to each program. Changes were based on the continuing low demand for U.S. timber (noted a USDA budget analyst: OMB "appears to be skeptical of their own economic assumption of a high level of housing starts."[19]) and a view, not shared by the U.S. timber industry, that the Canadian import situation should be resolved largely by the free market.

In a section of the OMB passback entitled "Director's Review Decisions," supplementing the recommendations of the career cadre at OMB, the Executive Office of the President makes clear its own agenda for the Forest Service.

* All State and Private Forestry assistance grants are to be terminated after FY 1985.

* Plans for BLM/Forest Service jurisdictional restructuring (the so-called "interchange" proposed by BLM Director Robert Burford, an administration appointee) are to be implemented in FY 1986. Administration of Oregon and California grant lands is to be transferred from BLM to Forest Service during FY 86.

* Grazing fees based on fair market value are to be determined at NFS regional levels and phased in over a period of three years.

* An administration legislative proposal is to be drafted to increase recreation user fees to recover 25 percent of the total costs of providing recreation on the national forests.

December 1984: The funding reductions proposed by OMB seem to run absolutely counter to the policies and priorities of Assistant Secretary Crowell for management of the National Forest System. Procedure dictates that any appeal of the OMB proposal to OMB Associate Director Khedouri has to be initiated by Agriculture Deputy Secretary Richard Lyng, who is too preoccupied at the time with an array of more mainstream agricultural policy questions coming before Congress (the farm bill, commodity price supports, a Soviet grain export treaty) to focus attention on the national forest timber sale level. The remarkable degree of independence and autonomy the Forest Service has always enjoyed within USDA may work against the agency in the context of the budget process. Appeals of OMB decisions must be made to OMB by the deputy secretary of agriculture. If the deputy secretary knows little about Forest Service needs, or is preoccupied with mainstream agricultural policy issues, then the Forest Service essentially has no high level political representation in this important stage of the budgetary process.

For FY 1986, Assistant Secretary Crowell failed to enlist the aid of Deputy Secretary Lyng and was left to petition OMB without the leadership of a high-level administration appointee. Associate Director Khedouri, on a political plane higher than that of an assistant secretary, declined to respond to Crowell until shortly before the President's Budget was due to go to press (Crowell had refused to release the final USDA budget estimates to OMB for preparation of the President's Budget until OMB agreed to review his concerns). This issue, among others, was finally taken up by the OMB Review Board established to hear the appeals of cabinet agencies (and consisting of White House Counsel Edwin Meese, White House Chief of Staff James Baker, OMB Director David Stockman, and Agriculture Secretary John Block) where the decision was made to marginally raise the timber sale level to 10.7 BBF. Crowell was informed of the decision as the President's Budget was about to be released when nothing further could be done to change the figures. A summary of executive branch actions on the FY 1986 budget proposal for the Forest Service appears in Table 3.1.

Chronology of Congressional Action on, and Related to, Forest Service Appropriations for FY 1986

At various times during its passage through the Congress, the Forest Service budget request is reviewed by no less than ten committees and subcommittees plus the full House, full Senate and a House/Senate conference committee. The process specified by the Congressional Budget and Impoundment Control Act of 1974, which was the process followed for FY 1986 appropriations, is illustrated in Figure 3.2. This process was modified for FY 1987 and afterwards by the Balanced Budget and Emergency Deficit Control Act of 1985 (Gramm-Rudman-Hollings)--this modified process is illustrated in Figure 3.3. If everything goes according to the schedule, this congressional portion of the budget process takes a full nine months. If things do not go according to schedule, which typically is the case, it may take until mid-December for the Congress to come to agreement and pass an appropriations bill for the Forest Service. In other cases, the Congress may not ever come to agreement on a particular appropriations bill (there are thirteen separate appropriations bills to span all of the functions of the federal government) and all or several of the appropriations bills may be folded into an "omnibus budget resolution" which, due to the impending urgency of funding most of the operations of the federal government, is seldom resisted by members of Congress even though they may still disagree with some portion of the resolution as it affects a particular agency or program.

By convention, although not by constitutional authority, all appropriations bills originate in the House.[20] After passage by the full House they are responded to by the corresponding Senate committees and full body, and any differences are then ironed out in a House-Senate conference committee chaired alternately by a House member or Senator. The first panel to review the President's Budget for the Forest Service is the House Appropriations Subcommittee on Interior and Related Agencies, which also has responsibility for funding all of the agencies in the Department of the Interior, from the national parks to the National Endowment for the Arts, to natural gas leasing on the Outer Continental Shelf. Hearings are held simultaneously by the House Budget Committee, which has no subcommittees specific to agencies or programs, to develop overall spending limits that the appropriations committees are obliged to stay within. The Appropriations Committee offers preliminary recommendations to the Budget Committee in the form of "Views and Estimates of Appropriation Plans." The standing committees, in this case the House

Table 3.1
Summary of Executive Branch Actions on the Forest Service Budget for Fiscal Year 1986

RESOURCE PROGRAM	BASE	(A) AGENCY REQUEST	USDA ALLOWANCE	CHANGE FROM (A)	(B) PRESIDENT'S BUDGET	CHANGE FROM (A)
	[M$]	[M$]	[M$]		[M$]	
TIMBER	990,308	1,009,460	1,080,346	7.0%	903,590	-10.5%
RECREATION	130,217	138,041	116,605	-15.5%	115,717	-16.2%
FISH & WILDLIFE	65,938	56,887	55,538	-2.4%	55,130	-3.1%
SOIL & WATER	53,035	45,633	43,810	-4.0%	44,442	-2.6%
RANGE	32,136	34,698	32,378	-6.7%	32,149	-7.3%
MINERALS	26,572	29,539	29,689	0.5%	29,379	-0.5%
PROTECTION	156,591	165,379	165,993	0.4%	147,507	-10.8%
LANDS	49,926	54,156	53,236	-1.7%	50,129	-7.4%
FACILITIES	117,506	133,196	127,630	-4.2%	112,728	-15.4%
TOTAL	1,622,230	1,666,988	1,705,225	2.3%	1,490,771	-10.6%

All funding figures are in thousands of constant 1982 dollars.

Figure 3.2
The Congressional Budget Process Under the Congressional Budget and Impoundment Control Act of 1974

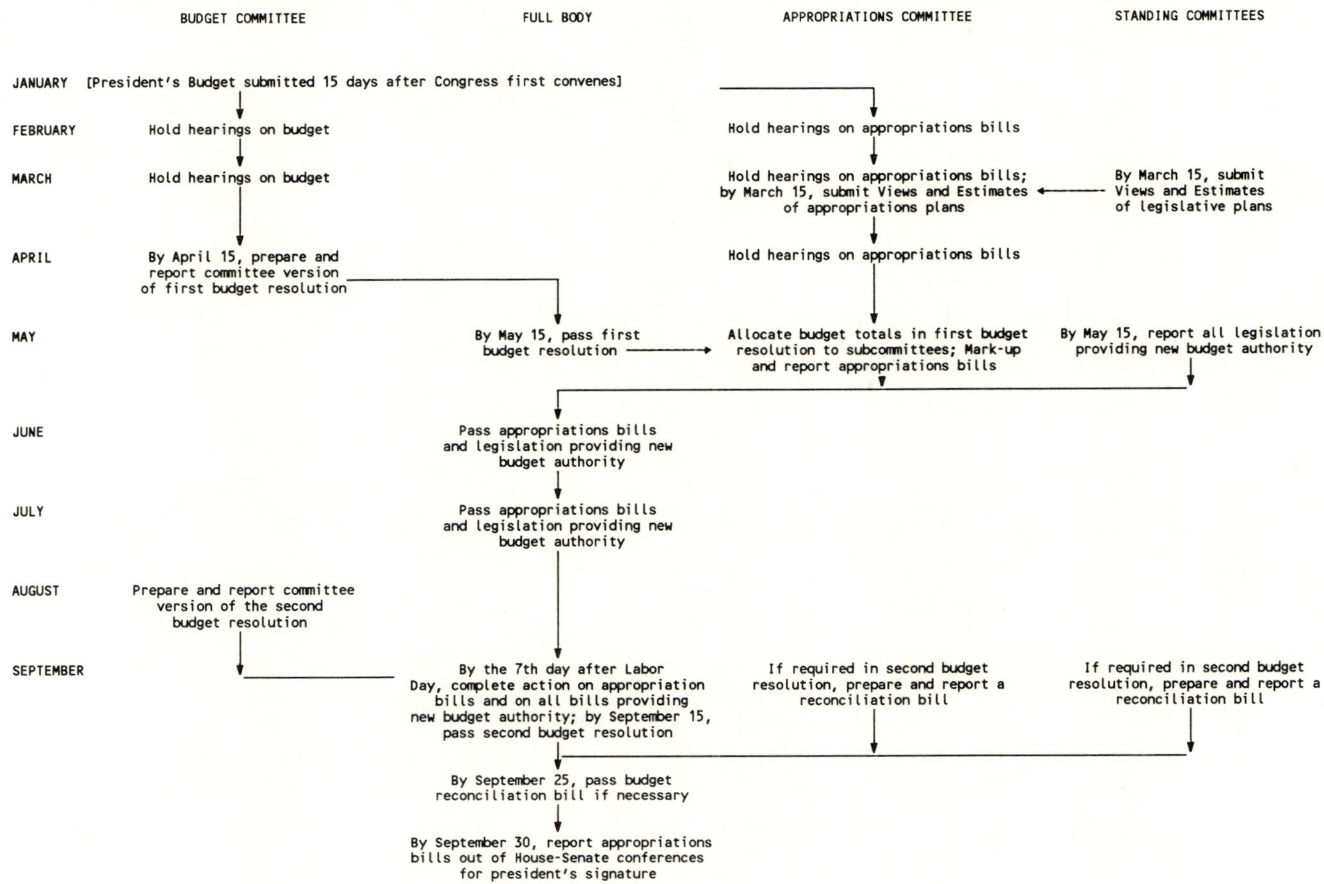

Figure 3.3
The Congressional Budget Process Under the Balanced Budget and Emergency Deficit Control Act of 1985 ("Gramm-Rudman-Hollings")

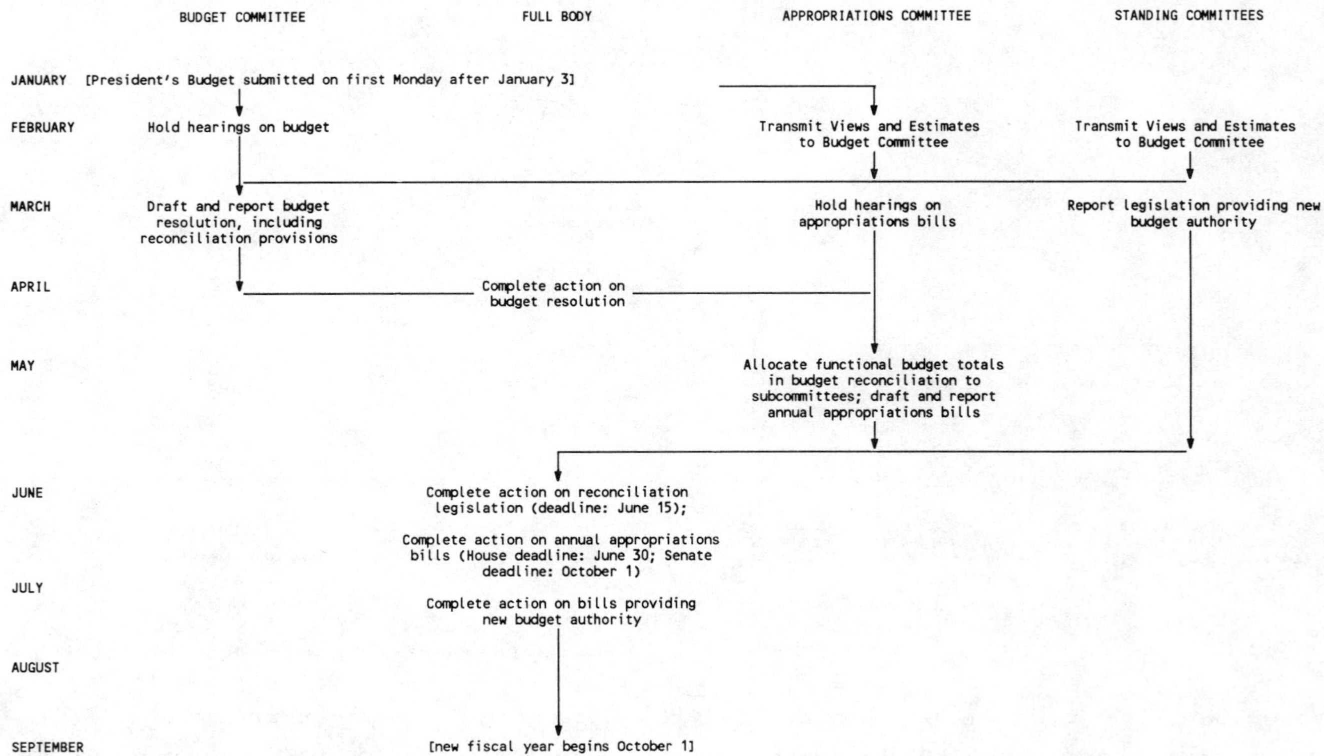

Interior and Insular Affairs Committee (Subcommittee on Public Lands and National Parks) and the House Agriculture Committee (Subcommittee on Forests, Family Farms and Energy), review the budget request and also send their recommendations to the Budget Committee as "Views and Estimates of Legislative Plans." House rules prohibit legislating on an appropriations bill, defined generally as specifying a course of action to be followed by an agency for more than just the upcoming fiscal year. If a standing committee feels that an issue being taken up by the Appropriations Committee involves broader policy issues, this rule can be invoked so that the issue can be dealt with more fully in hearings and legislation originating with the standing committee.

Except for a few budget line items at the center of major controversies, such as the disputes over the timber harvest level or the level of road construction, or of special interest to one or more legislators, the appropriations for many line items are virtually the same as proposed in the President's Budget. As shown in the summary of congressional actions on the FY 1986 Forest Service budget (see Table 3.2), the House made a major reduction (-28%) in the timber program, a major increase (15%) in the recreation program and relatively minor changes for all other resource programs. Generally, the House acted to reduce the OMB proposed funding for commodity resource programs (timber, range and minerals), and increase OMB's allowances for the non-commodity resource programs (recreation, fish and wildlife, soil and water and fire protection). The House also reduced funding for the lands program and for facilities construction and maintenance, two programs for which funding had already been significantly reduced by OMB itself. Responding to the unfolding timber industry recession, the Senate, too, reduced funding for timber sales, although not by as much as the House. The recreation program was given a modest increase, primarily to boost funding for recreation support activities associated with timber harvesting, and more substantial increases were made in the fish and wildlife program and the fire protection program.[21] The remaining programs were funded almost exactly as in the President's Budget. In the final appropriation, the House and Senate split the difference between them, with the majority of the resource programs being funded within 4 percent of the level proposed by OMB.

Within the timber program, the entire debate between the House and Senate was focused on three segments: timber sales, road construction and reforestation. Where the House cut 14 percent from timber sales in the President's Budget, the Senate added it all back, plus 1.6 percent. Where the House cut $50 million for national forest road construction, 23 percent of the total, the Senate restored funding to the level in the President's Budget. The differences between the two chambers on reforestation were largely the result of an accounting change, with $30 million being transferred from the Reforestation Trust Fund to the regular National Forest System reforestation account

Table 3.2
Summary of Congressional Changes to the FY 1986 Forest Service Budget

RESOURCE PROGRAM	PRESIDENT'S BUDGET (OMB)	HOUSE	CHANGE FROM FS REQUEST	CHANGE FROM OMB	SENATE	CHANGE FROM FS REQUEST	CHANGE FROM OMB	FINAL APPROP	CHANGE FROM FS REQUEST	CHANGE FROM OMB
	[M$]	[M$]			[M$]			[M$]		
TIMBER	903,590	648,187	-35.8%	-28.3%	735,612	-27.1%	-18.6%	821,524	-18.6%	-9.1%
RECREATION	115,717	133,339	-3.4%	15.2%	118,617	-14.1%	2.5%	131,206	-5.0%	13.4%
FISH & WILDLIFE	55,130	55,341	-2.7%	0.4%	59,130	3.9%	7.3%	56,815	-0.1%	3.1%
SOIL & WATER	44,442	46,332	1.5%	4.3%	44,442	-2.6%	0.0%	45,014	-1.4%	1.3%
RANGE	32,149	30,426	-12.3%	-5.4%	32,649	-5.9%	1.6%	31,858	-8.2%	-0.9%
MINERALS	29,379	28,373	-3.9%	-3.4%	29,379	-0.5%	0.0%	28,341	-4.1%	-3.5%
PROTECTION	147,507	154,612	-6.5%	4.8%	160,291	-3.1%	8.7%	158,240	-4.3%	7.3%
LANDS	50,129	46,090	-14.9%	-8.1%	49,829	-8.0%	-0.6%	49,430	-8.7%	-1.4%
FACILITIES	112,728	105,223	-21.0%	-6.7%	115,528	-13.3%	2.5%	114,197	-14.3%	1.3%
TOTAL	1,490,771	1,247,923	-25.1%	-16.3%	1,345,477	-19.3%	-9.7%	1,436,625	-13.8%	-3.6%

All funding figures are in constant 1982 dollars.

by the Senate. It is notable that for a majority of the line items in the timber program, there was literally not a dollar's worth of difference between the OMB, House and Senate proposals.

The apparent increase in final timber appropriations over the allowances by both the House and the Senate is largely accounted for by the fact that the authorization for "purchaser road credits," a device by which the Forest Service can contract for the construction of timber roads by paying the contractor in federal timber credits rather than paying directly with appropriated dollars, was left out of the House and Senate committee reports but included in the conference report. This authorization is considered "off budget" because, it is argued, it does not involve the expenditure of appropriated funds and thus should not be subject to funding limitations under the budget reconciliation procedure. Because it is a credit given toward the purchase of federal timber, however, it is tantamount to an expenditure because it directly reduces federal receipts--by as much as $250 million annually. For FY 1986, the conference committee essentially authorized a $108 million reduction in receipts through the purchaser road credit system.

Within the recreation program, there seemed to be less attention focused on the management aspects of the program, recreation use management and trail maintenance, than on the facilities construction aspects. For recreation use management, the largest account, the House allowed slightly less than requested by the Forest Service but slightly more than proposed by OMB. The Senate attempted to cut it back to the OMB level but ultimately conceded almost completely to the House proposal in conference. Conversely, the Senate proposed to add to the House proposal for trail maintenance, topping the Forest Service request by 1.3 percent and exceeding OMB's proposal by nearly 11 percent; House and Senate split the difference between their respective proposals precisely down the middle and voted a 5.9 percent increase in the conference committee. Funding requested by the Forest Service for recreation construction, hit hard by both USDA and OMB, was largely restored in the House then hit hard again in the Senate. Although the final appropriation was 38 percent above what OMB proposed, it was still down 31 percent from the agency request and nearly $1 million below the prior year's appropriation. The Forest Service's trail construction request went unnoticed by USDA but was nearly halved by OMB. The House responded by pushing funding to nearly one-third more than requested by the Forest Service, 140 percent over the OMB level. The Senate maintained the OMB level unchanged but conceded in conference to more than double that. This pattern of behavior is consistent with the general pattern found over many past years: After significant OMB reductions in the recreation program, especially recreation construction, the House restores much of the funding and is able to sustain most of the increases in its conference with the Senate, which tends to hew close to the line established by OMB.

This is not the case for the fish and wildlife program. For FY 1986, the House essentially sustained modest OMB reductions in the Forest Service request. The Senate asked for an 11 percent increase and settled on half that in conference. With respect to the previous 25 years of Forest Service budgets, the more typical pattern is for the House and Senate to play the reverse roles. Authorization for expenditures from the Knutson-Vandenburg trust fund, which pays for a substantial share of the costs for wildlife habitat improvement done in association with timber management, remained virtually unchanged by all participants in the budget process. The same can be said for soil and water improvement activities funded through the Knutson-Vandenburg funds. Direct appropriations for the soil and water program followed its typical pattern of increase in the House, decrease in the Senate (i.e., no change from the reductions called for by OMB) and splitting the difference in the conference committee. As commodity resource programs, range and minerals shared the same general pattern of the House reducing the President's Budget, the Senate sustaining or increasing levels in the President's Budget and final appropriations nearly matching the President's Budget.

The facilities and lands programs displayed a pattern that is typical for these programs in particular--they came in for hard knocks from nearly every participant in the budget process. The real estate management portion of the lands program, which includes the administration of special use permits, was cut back nearly 13 percent by OMB, a reduction that was sustained through to the final appropriation. Every segment of the facilities program was significantly reduced by OMB, but none as severely as the construction of fire, administrative and other (FA&O) facilities (-48%). Though the House made no attempt to change this, the Senate moderated the reduction to only 37 percent, which approximated the final appropriation.

For the vast majority of National Forest System programs, then, the thrust and parry of the House and Senate seemed to largely cancel out one another, leaving a national forest budget very close to that proposed by OMB in the President's Budget. Most of the initiatives to change the course of existing policies came from the House, and the Senate acted to restate the President's Budget and resist any changes therefrom in conference. In the few cases of major controversies, such as over the House's substantial cuts in funding for timber sales and road construction, the Senate varied from its normal pattern (as illustrated in the time-series data above) of offsetting large cuts (or increases) in the House with correspondingly large increases (or cuts) in their own proposal, so that in conference it would be possible to compromise at somewhere near the President's Budget. Rather, the Senate proposal merely reasserted the OMB proposal. Senate appropriations staff members point out that while the results may suggest that the Senate acts primarily to moderate House changes to the President's Budget, its agenda is an independent one and is not set entirely by either the House or the president.[22] The House's special power in the appropriations process, the momentum it can establish as the originator of appropriations bills, seems to be tempered by the Senate's special power as an appeals court, as it were, for parties whose interests have been threatened by House actions. Once the House passes its bill and the bill is taken up by the Senate, the House's full hand is on the table; parties aggrieved by the House's actions can mobilize for revisions in two subsequent forums, the Senate and the conference committee.

It is important, for at least two reasons, to understand how the current structure of the appropriations process allows the Forest Service budget to be swept up in major spending controversies where the amounts in question utterly dwarf the entire Forest Service budget. The first is, again, the tactical advantage this affords the managers of the Senate Interior appropriations bill; the significance of Forest Service appropriations in an appropriations bill that includes funding for all programs in the Department of the Interior and nearly all the nonweapons programs of the Department of Energy (the Forest Service budget is usually less than 20 percent of the overall Interior appropriations bill) relegates them to a lesser role when it comes to meeting overall spending limitations within the Senate's budget reconciliation process. The only case in which the reconciliation process would materially affect Forest Service appropriations is when an across-the-board reduction is specified, as was threatened for the FY 1986 bill by Senate Budget Committee Chairman Domenici. This minor role becomes a powerful tool for national forest interests in the Senate, which can use the momentum of the full appropriations bill to overwhelm House opposition on specific policy disputes. And when the option becomes available to Senate tacticians, the use of a single, massive, year-long continuing appropriation in place of individual appropriations bills provides an extreme example of this. When Forest Service appropriations are rolled into a continuing resolution that holds the spending authority for most of the federal government's operations, and the fiscal year has already begun, there is overwhelming pressure on the House to relinquish its positions on specific policy disputes in order to conclude the overall process so that the government can continue to function uninterrupted.

The second reason is to better understand how the appropriations process, as it currently functions, affects the timing of Forest Service appropriations and thereby the execution of the budget provided by Congress for the Forest Service. The House passed its Interior appropriations bill on July 31, 1985 and the full appropriations committee of the Senate reported the bill to the full Senate on September 24. Due to larger factors with almost no relevance to Forest Service programs, such as whether to continue funding the Synthetic Fuels Corporation and whether to maintain a moratorium on oil and gas leasing on the Outer Continental Shelf off California, an Interior appropriations bill as such was never passed, but had to be rolled into a massive continuing resolution comprising appropriations for nearly every department in the federal government outside of Defense and State. And it was not until December 19, nearly three months into fiscal year 1986, that the Forest Service could be sure of just what Congress' spending limits and priorities would be for its programs. As will be seen in the following section on the execution of the FY 1986 Forest Service budget, this delay--and the fact that such delays have become almost institutionalized--has had a significant impact on how the Forest Service plans and carries out its annual management activities on the national forests.

NOTES

1. A more detailed description of the unique history of the eastern national forests, and an outstanding examination of the opportunities and challenges facing the managers of these resources, can be found in: Shands, W., and Healy, R., *The Lands Nobody Wanted: Policy for the National Forests of the Eastern United States* (Washington, D.C.: The Conservation Foundation, 1977).

2. deBuys, William, *Enchantment and Exploitation: The Life and Times of a New Mexico Mountain Range* (Albuquerque, N.M.: University of New Mexico Press, 1985).

3. The "administrative acreage" of the Mt. Baker-Snoqualmie National Forest is approximately 1.7 million acres. The boundaries include about 400,000 acres of private and state land inholdings; another 800,000 acres on the east side of the Cascades is actually administered as part of the Wenatchee National Forest.

4. USDA Forest Service, *FY 1986 Program Budget Instructions*, unpublished internal memorandum, Washington Office, September 16, 1983, p. 5.

5. This set of events is extremely well documented in: McKenney, J., and Cale, E., *An Analysis of the 1979 PPB Process of the U.S. Forest Service*, Unpublished report available from the author, Harvard Graduate School of Business Administration, Soldier's Field, MA, November 1977.

6. USDA Forest Service, *FY 1986 Program Budget Instructions*, p. 5.

7. USDA Forest Service, *Report of the Forest Service for Fiscal Year 1983* (Washington, D.C.: GPO, 1984).

8. Kaufman, Herbert, *The Forest Ranger: A Study in Administrative Behavior* (Baltimore: Johns Hopkins University Press for Resources for the Future, 1967), p. 83.

9. Memorandum from Planning, Programming and Budgeting, R9, to Forest Supervisors and Staff Directors, September 12, 1983.

10. Ibid.

11. Memorandum for Regional Forester, R9, to Forest Supervisors and Staff Directors, October 18, 1983.

12. Ibid.

13. USDA Office of Budget and Policy Analysis, internal staff report, November 1984.

14. The Federal Timber Contract Modification Act, P.L. 98-478.

15. USDA Office of Budget and Policy Analysis, "Decision Paper for Deputy Secretary's Mark-Up," Internal Memorandum, November 1984.

16. Ibid., p. 2.

17. Ibid.

18. Office of Management and Budget, "FY 1986 OMB Passback for Department of Agriculture," Internal Interdepartmental Memorandum, November 1984.

19. USDA Office of Budget and Policy Analysis, "FY 1986 Forest Service Budget, November Situation," Internal Staff Report, November 1984, p. 3.

20. This practice grew out of early U.S. history, when the House Ways and Means Committee, whose special jurisdiction is taxation, also appropriated funds. Article I of the Constitution states that "all Bills for raising Revenue shall originate in the House of Representatives" but makes no such provision for appropriations. This ambiguity was the basis for a famous battle in 1962 between the chairmen of the House Appropriations Committee and the Senate Appropriations Committee, the resolution of which was an agreement whereby the House would continue to originate all appropriations bills, for which it has no constitutional authority, and permit all tax bills, for which the House has clear constitutional authority, to go to the Senate. See: Shuman, H. E., *Politics and the Budget: The Struggle Between the President and the Congress* (Englewood Cliffs, N.J.: Prentice-Hall, 1984), pp. 65-67.

21. U.S. Senate, Committee on Appropriations, *Appropriations for the Department of the Interior and Related Agencies for Fiscal Year 1986*, Senate Report 99-141 (Washington, D.C.: GPO, September 24, 1985).

22. Personal communication, Senate Interior Appropriations Subcommittee staff.

4

Conceptual Models for Explaining Modifications Made to the Forest Service Budget by External Organizations

In a sense, the proposals by each budget process participant for a different distribution of funding among the various NFS resource programs, as illustrated in Table 2.4, can be thought of as statements of different priorities between the participants. For each participant, there is a different optimal distribution of funding that best represents the values and goals of that organization for the management of national forest resources. These priorities are all the more eloquently revealed in the way each participant's proposed distribution shifted as the overall Forest Service budget went from the growth period of FY 1977-1981 to a period of significant budget contraction in FY 1982-1986. As the overall budget declines, programs of higher priority constitute an increasing share of the overall budget proposal as the participant seeks to sustain those programs at or near constant levels. Programs of lower priority are thus forced to decline even more rapidly than the budget overall. As shown in the table, each of the executive branch participants give somewhat higher priority to commodity programs such as timber and minerals, varying only in the degree of that emphasis; Congress seems to favor a more "across-the-board" approach that maintains the same relative emphasis among programs across a fairly wide range of overall agency funding levels. Almost without exception, the foregoing analysis of each participant's modifications of the Forest Service's annual budget requests over many years shows their actions to be consistent with the emerging pattern of what might be termed their revealed priorities.

There is a tendency when viewing the broad pattern of budget outcomes to see an overarching rationality to this progression of events, leading to the government's benevolent and omniscient allocation of its monetary resources to nurture and cultivate our national forests, and at the same time build our bombers and battleships and sustain our cities and farms. Forest Service budget handbooks themselves encourage the notion, dutifully conveyed by many field officials to the public with an air of perennially renewed resolve, that our elected government and representatives, in their collective wisdom, have spoke. It is then the function of the Forest Service officials to simply execute the tasks, specific and implied, that have been assigned to them through the annual budget. This "Rational Policy Model," as described by Allison, assumes that "if a nation performed a particular action, that nation must have had ends toward which the action constituted an optimal means."[1] In this case, the end toward which the government as a whole was working may have been something like "reduce federal expenditures to eliminate the budget deficit while protecting the full range of resources on the national forests and, where possible, enhancing federal revenues through the sale of national forest commodity resources;" or it may have been some other goal yet to be articulated. Nevertheless, the assumption is that the decisions that emerge from the process are "purposive acts of a unified government . . . actions chosen by a unified, rational decision maker: centrally controlled, completely informed and value maximizing;" to understand the rationale for these acts one needs only to discover what the overarching goal is or was.[2]

Though such goals as budget deficit reduction may indeed serve as an integral consideration in the decisions of each budget participant, particularly when this problem is as acute as it was in the mid-1980s, one needs only to pick up a newspaper to know that the budget actions of the executive and legislative branches of the federal government are hardly unified, centrally controlled or even completely informed. However, it is much less obvious to most observers that the budget actions within the executive branch, which culminate in the President's Budget, do not fit the Rational Policy Model very well. Perhaps in part because of the degree to which attention is focused on budget and policy differences between the Congress and the current administration, many see policy decisions emanating from the executive branch, even those on agency budgets, as the product of a cooperative and monolithic effort. For its part, OMB promotes and encourages this view through its restrictions on the release of any information pertaining to budget proposals that are preliminary to the President's Budget. In effect, the administration speaks with one voice and very few individuals, often not even members of Congress, are privy to policy or program disagreements that may have existed among the different organizations and individuals within the executive branch. Unless information is deliberately leaked, the budget process inside the executive branch is performed in secret. Members of Congress may know that decisions are being made that will act to limit their own flexibility when the President's Budget is eventually submitted to them, but often are frustrated in their attempts to learn the nature of internal policy disputes and how they will affect the subset of the total information that will subsequently be made available to them. The late Senator Hubert Humphrey once noted that:

> Whether there is a Republican or Democrat in the White House, the budget has remained more secret than the plans for an atomic bomb. . . . As the proposals move from the various departments and agencies through [OMB] to the President, it is a subliminal process. Simply, the bureaucracy can be an enemy of what is proper and necessary in a federal system.[3]

The inscrutability of the process is only deepened by OMB's further restrictions on the public release of information on internal executive branch budget negotiations even after appropriations for the fiscal year have been finalized. After the fiscal year is over, little of the information can be released without a specific request under the Freedom of Information Act and, even at that, most of the information has been declared exempt from FOIA and therefore can never be officially released. Any signs of internal conflict are hidden from the outside observer; "intense political activity is skillfully and scrupulously camouflaged so that the resulting policies appear to be in harmony with the official ideology."[4]

As the foregoing analysis of budget decisions and negotiations clearly indicates, the executive branch is neither monolithic nor unified. In reality, "a government consists of a conglomerate of semi-feudal, loosely allied organizations, each with a substantial life of its own."[5] Among the organizations associated with the Forest Service budget there is specialization both horizontally and vertically. It is the Forest Service's task to protect and manage the national forests; OMB seeks to maximize the efficiency of government spending within overall budget constraints; the secretary of agriculture attempts to apply the general philosophies and policies articulated by the president to an array of agricultural programs, only one of which involves the national forests.

> To be responsive to a wide spectrum of problems, governments consist of large organizations, among which primary responsibility for particular tasks is divided. Each organization attends to a special set of problems and acts in quasi-independence on these problems. But few important issues fall exclusively within the domain of a single organization. Thus, government behavior relevant to any important problem reflects the independent output of several organizations, partially coordinated by government leaders. Government leaders can substantially disturb, but cannot substantially control, the behavior of these organizations.[6]

As a function of its enabling or organic legislative mandate and subsequent evolution in response to the tasks it is faced with, each organization within the government develops a set of standard operating procedures to deal with both routine duties and reasonably predictable contingencies. The larger the organization and the more complex its duties, the greater the need for standard operating procedures to coordinate and control the actions of individuals at the operating levels of the organization. In the case of the Forest Service, the coordination and control task is complicated by a highly decentralized organizational structure, comprising more than 40,000 employees scattered in nearly a thousand different locations from Puerto Rico to Alaska.[7] While these standard operating procedures enable the organization to fulfill its responsibilities, they also come to define the organization's capabilities and, through their sheer momentum, make it difficult for the organization to quickly redirect itself in response to an unexpected contingency or a rapidly changing environment.

An understanding of an organization's standard operating procedures, and the organizational history, values, traditions and incentives that underlie them, can serve as a reliable basis for predicting such "lower functions" as budget preparation and even such "higher functions" as how the organizations' top officials will perform the routine of negotiating budgets. Such an understanding improves the ability of the other involved organizations to minimize uncertainty and continue in their tasks with as little change as possible. This is the basis for a second conceptual model for exploring the rationale behind a particular action or decision by a government or any other large organization. In what Allison terms the "Organizational Process Model," actions can be understood "less as deliberate choices and more as *outputs* of large organizations functioning according to standard patterns of behavior."[8] The rationale for the action can be explained through an exploration of the standard operating procedures and organizational routines within the organization that produced the action.

There is, in fact, a continuing concern among democratic theorists that the decisions of government are not "deliberate choices" by elected government representatives as much as outputs of the bureaucracy.[9] The specialization of organizations within the government and the increasing complexity of the government's tasks and responsibilities vest an increasing share of the executive branch not in the elected leaders at the top, but in the agencies and departments. Understanding and managing the agency's responsibilities requires greater and greater specialization and technical expertise. And increasingly, the agencies themselves are the only ones with the time and interest to develop this expertise. An agency has vastly more time and personnel to devote to a policy issue than political leaders, who can pay only sporadic attention to an issue that an agency may have worked on for months or years. Having a 40,000-person Forest Service allows much greater attention to managing the national forests than if the task were to be done by the staff of the Office of the Secretary of Agriculture, especially given the other tasks assigned to the department. Likewise, having a Department of Agriculture permits a more effective management of agricultural programs, including management of the national forests, than if the task were taken on directly by OMB or even by the Executive Office of the President as a whole. But the responsibility that is handed down the hierarchy carries with it a large measure of discretion and control over which responsibilities are tended to and how. Though arranged hierarchically and acting sequentially on the Forest Service budget, neither the secretary of agriculture nor OMB effectively holds command authority over the Forest Service. With the specialization of organizations within the government by task there comes a vertical sharing of power among those organizations.

As the complexity of responsibilities increases and the differential in expertise continues to widen, so grows the agency's control over its own policy outcomes. Particularly in the case of a technically oriented agency such as the Forest Service, a near-monopoly on information is gained and political leaders must rely increasingly on the information the agency chooses to provide them. Although political leaders may make the final decision, the agency is in a position to influence this decision by shaping the relevant information and defining the options. By packaging the information, the agency can steer political decision makers so that they choose a course of action roughly in line with what the agency wants. What information is packaged, and how, is a key organizational output. The structure of the situation may be so dominated by these organizational outputs that the formal decision by the head of the government is anticlimactic. Not having the agency's specialized knowledge and expertise, the decision maker often has no way to recognize that there may be other viable alternatives that do not appear on the agenda--and thus cannot be selected.[10] Former White House Counsel Theodore Sorenson once noted that even presidents

> rarely, if ever, make decisions in the sense of writing their conclusions on a clean slate. . . . The basic decisions, which confine their choices, have all too often been previously made.[11]

The power of the Forest Service as a large and complex expert organization is tempered by the fact that specialization, and thus power sharing, within a government is horizontal as well as vertical. The Organizational Process Model may convey a deeper understanding for the actions, or outputs, of a single, distinct organization within the government, such as the Forest Service. But it falls short of explaining the ultimate decision emanating from the organization as a whole, the executive branch, when several powerful organizations with overlapping interests, and with histories and values as distinctly different as the Forest Service's and OMB's, collide in the decision making process that produces the singular outcome we see as the

President's Budget. Conflicting goals and interests within a hierarchy are, in themselves, not particularly interesting; but a structure of shared power among several quasi-independent organizations with overlapping interests presents a much more complex environment in which any outcome from the overall organization must be the product of competition, bargaining and strategy. Drawing from the evolution of the concept by Truman[12] Dahl[13] and Lindblom,[14] Hilsman asserted that a political process exhibits three distinct characteristics:

1. there is a diversity of goals and values that must be reconciled before a decision can be reached;

2. there are competing clusters of people within the main group who are identified with each of the alternative goals and policies; and

3. the relative power of these different groups of people included is as relevant to the final decision as the appeal of the goals they seek the cogency and wisdom of their arguments.[15]

Thus, in a third conceptual model, later termed by Allison the "Bureaucratic Politics Model," the decision or action that ultimately emerges is the consequence of

> innumerable and often conflicting smaller actions by individuals at various levels of bureaucratic organizations in the service of a variety of only partially compatible conceptions of national goals, organizational goals and political objectives.[16]

A critical additional consideration in this conceptual model is that rather than a rational policy choice of a unitary actor (Rational Policy Model) or even the rational output of organizations with distinct and identifiable goals and values (Organizational Process Model), the Bureaucratic Politics Model sees the action or decision that emerges as an *outcome* of different groups pulling in different directions and producing a result that may be quite different from what any one of the groups intended. The resulting action can be thought of as the political equivalent of a vector product, a course whose direction is a function not only of directions from which the various forces act upon the decision maker, but the relative strengths of those forces as well. The advocate of a particular course of action must build consensus to support that course of action; "where there are rival advocates or rival policies, there is competition for support, and all the techniques of alliance appear--persuasion, accommodation, and bargaining."[17]

Organizations, as such, do not bargain--people do. An exploration of the interactions between organizations leading to a particular outcome must therefore take into account the personalities, values and tactics of individual people, the central players for each organization who interact with one another to represent their organizations and themselves.

> The core of the bureaucratic politics mix is personality--each player's basic operating style, the complementarity or contradiction among personalities and styles in the inner circles are irreducible pieces of the policy blend. Moreover, each person comes to his positions with baggage in tow, including sensitivities to certain issues, commitments to various programs and personal standing and debts with groups in the society.[18]

Analysis of policy outcomes in terms of the roles and actions of individual players poses a new dimension of complexity. Seldom is the interplay of the central actors in the "inner circles" of federal governmental policymaking committed to the written record. Indeed, in the case of budget policymaking within the executive branch, this interplay is explicitly *not* a matter of public record. Nonetheless, valuable additional insights into this aspect of the budgeting process can be gained from what unofficial written records do exist, and from the personal recollections of agency and other governmental personnel directly and indirectly associated with the discussions and negotiations.

In this analysis of the processes by which the budgets for national forest management activities have been shaped, these three conceptual models (Rational Policy, Organizational Process and Bureaucratic Politics) are not employed independently but in concert. They can be thought of as progressively more powerful lenses on the same microscope, each one revealing different aspects and greater detail of the dynamic organism under study, and providing a fuller understanding of how the overall organism functions. It is important to recognize that this analysis is an attempt to describe and explain policy outcomes rather than an attempt to derive a model for predicting or prescribing governmental behavior or actions. The academic literature contains a variety of parsimonious mathematical models that attempt to predict or prescribe organizational actions, especially budgetary decision making by governments.[19] The political models discussed here, the second and third conceptual models, make no such attempt at parsimony or prediction, but rather attempt to explicitly recognize the enormous number of variables extant in decision making. An observation made by Allison in 1971 is still current: "The gap between academic literature and the experience of participants in government is nowhere wider than at this point."[20]

On the other hand, an exclusive reliance on the political models has two drawbacks.

1. The fact that power is shared both vertically and horizontally within the executive branch does not obviate the need to consider that some portion of the conflict among agencies and departments is governed, in accordance with classical Weberian theory, by the hierarchy of the governmental structure.[21] To a large degree, the conflicts that arise are generated, and resolved, with respect to this hierarchy.[22] Decision analysis cannot be factored out of the political model; the political model simply offers deeper insights in the study of conflict resolution using decision analysis.

2. An overly narrow use of the political models, even in combination with traditional decision analysis, focuses exclusively on intra-organizational conflict and does not look at the organization in terms of its environment. One cannot ignore the effects of events external to the organization and the response of the organization to these events. External events affecting conflict generation and resolution in the context of the Forest Service budget might include policy controversies bubbling up from the public or interest groups, or policy initiatives from the Congress. Organization for analysis purposes is a "social *system* composed of various subsystems linked together by mutually interdependent bonds."[23] This precaution applies to budget practitioners as well as students of the budget process; significant changes in the external environment are often ignored by the standardized and routinized budget process--issues that are not dealt with effectively by the agency often rise to the attention of Congress, stimulating a reaction that may find its expression in Congress' appropriations to the Forest Service.[24]

Before exploring the observed behavior and outcomes in terms of these political models, it will be useful to take a more detailed look at each of the organizations participating in the Forest Service budget development process. Through a brief examination of the historical development of each organization, and the values and incentive structures that have evolved from this history, it is possible to gain a deeper understanding not only of the standard operating procedures and organizational outputs that we can observe today, but of the fundamental forces underlying the observable phenomena and perhaps guiding their future evolution as well.

NOTES

1. Allison, Graham T., *Essence of Decision: Explaining the Cuban Missile Crisis* (Boston: Little, Brown and Company, 1971), p. 67.

2. Ibid.

3. Humphrey, Hubert H., *The Education of a Public Man* (Garden City, N.Y.: Doubleday, 1976), pp. 192-193.

4. Mouzelis, Nicos, *Organization and Bureaucracy* (Chicago: Aldine, 1967), p. 159.

5. Allison, Graham T., *Essence of Decision: Explaining the Cuban Missile Crisis.*

6. Ibid.

7. The classic study in how the Forest Service has dealt very effectively with this coordination and control task in a highly decentralized organization is Herbert Kaufmann's *The Forest Ranger: A Study in Administrative Behavior* (Baltimore: Johns Hopkins University Press for Resources for the Future, 1967).

8. Allison, Graham T., *Essence of Decision: Explaining the Cuban Missile Crisis.*

9. Etzioni-Halevy, Eva, *Bureaucracy and Democracy: A Political Dilemma* (Boston: Routledge and Kegan Paul, 1983); see also: Heclo, Hugh, *A Government of Strangers: Executive Politics in Washington* (Washington, D.C.: Brookings Institution, 1977).

10. Bachrach, Peter and Baratz, Morton, "The Two Faces of Power," *American Political Science Review*, Vol. 56 (1962) pp. 947-952. See also: Bachrach, Peter and Baratz, Morton, "Decisions and Non-decisions: An Analytical Framework," *American Political Science Review*, Vol. 57 (1963) pp. 632-642.

11. Sorenson, Theodore, "You Get to Walk to Work," *New York Times Magazine*, March 19, 1967, cited in: Allison, G., "Conceptual Modes and the Cuban Missile Crisis," *American Political Science Review*, Vol. 53 (1969) pp. 689-718 at p. 701.

12. Truman, David B., *The Governmental Process* (New York: Knopf, 1961).

13. Dahl, Robert, *A Preface to Democratic Theory* (Chicago: University of Chicago Press, 1956).

14. Lindblom, Charles, E., *The Intelligence of Democracy* (New York: Free Press, 1965). Lindblom, C. E., "The Science of Muddling Through," *Public Administration Review*, Vol. 19 (Spring 1959). Lindblom, C. E., "Still Muddling, Not Yet Through," *Public Administration Review*, Vol. 39 (1979). Lindblom, C. E. and Braybrooke, D., *A Strategy of Decision* (Glencoe, IL: 1963).

15. Hilsman, Roger, *To Move A Nation*, (New York: 1967), pp. 554-555.

16. Allison, Graham T., "Conceptual Modes and the Cuban Missile Crisis," p. 691.

17. Hilsman, *To Move A Nation*, p. 561.

18. Allison, "Conceptual Models and the Cuban Missile Crisis," p. 714.

19. See: Davis, Otto, Dempster, M. A. and Wildavsky, Aaron, "A Theory of the Budgetary Process," *American Political Science Review*, Vol. 40, No. 3 (1966), pp. 529-547; Natchez, Peter B. and Bupp, Irvin C., "Policy and Priority in the Budgetary Process," *American Political Science Review*, Vol. 67, No. 12 (1973), pp. 951-963; Padgett, John F., "Bounded Rationality in Budgetary Research," *American Political Science Review*, Vol. 74, No. 4 (1980), pp. 354-372; Bailey, John J., and O'Connor, Robert J., "Operationalizing Incrementalism: Measuring the Muddles," *Public Administration Review*, Vol. 35, No. 1 (January/February 1975), pp. 60-66; Crecine, John P., "A Computer Simulation Model of Municipal Budgeting," *Management Science*, Vol. 13, No. 7 (1967), pp. 786-815.

20. Allison, *Essence of Decision*, p. 146.

21. Weber, Max, *The Theory of Social and Economic Organization*, translated by A. M. Henderson and Talcott Parsons (New York: Oxford University Press, 1947), p. 331.

22. Blau, Peter M. and Meyer, Marshall W., *Bureaucracy in Modern Society* (New York: Random House, 1956), pp. 61-77.

23. Katz, Daniel and Kahn, Robert, *The Social Psychology of Organizations* (New York: Wiley, 1966). [emphasis added].

24. Crecine, John P. and Fisher, Gregory, *On Resource Allocation in the Department of Defense* (Ann Arbor, MI: Institute for Public Policy Studies, 1971), cited in: Thompson, James C., *Rolling Thunder: Understanding Policy and Program Failure* (Chapel Hill, N.C.: University of North Carolina Press, 1980), p. 114.

5

The Roles of Organizations Influencing the Forest Service Budget

In his classic treatise *The Politics of the Budgetary Process*, Aaron Wildavsky observed that the roles adopted by each of the major participants in the budget process, defined as "the expectations of behavior attached to institutional positions," set up patterns of mutual expectations that aid each participant in its strategizing and serves as a coordinating mechanism for the process overall.[1] In very general terms, Wildavsky offered the following definition of the roles of the major participants:

> the administrative agencies act as advocates of increased expenditure, [OMB] acts as presidential servant with a cutting bias, the House Appropriations Committee functions as a guardian of the Treasury, and the Senate Appropriations Committee as an appeals court to which agencies carry their disagreements with House action.[2]

The stability of these roles allows each participant to anticipate the actions of the other participants in strategizing to achieve a final outcome that will match closely with its particular goals.

> Strategies are the links between the goals of the agencies and their perceptions of the kinds of actions which will be effective in their political environment. . . . The roles of the major participants are powerful, persistent, and strongly grounded in the expectations of others as well as in the internal requirements of the positions.[3]

In terms of the conceptual models illustrated above, these roles are organizational outputs, functions of the standard operating procedures that have evolved over time in each organization. Following is a more detailed examination of the histories and incentive structures of each of the organizations that participate in the Forest Service budget process, intended to provide additional insights into *why* each organization played the role observed in the time-series data and in Forest Service budget development for FY 1986.

The Office of the Secretary of Agriculture

The function of the Office of the Secretary of Agriculture as an organization is to develop and execute agricultural policies and programs that directly reflect the general policy goals and philosophies espoused by the president. In so doing, the secretary's office is an explicitly political organization in which the goals and values are established anew with each presidential inauguration. Nearly all of the executives of the organization are, by definition, "outsiders" to the federal bureaucracy, entering government service for a brief and fixed period of time with the expectation of returning to private life at the end of that period. As with the political executives of every other department within the federal government, the effectiveness of the political executives in the Department of Agriculture in understanding and directing the activities of the agencies they oversee is limited by structural as well as personal factors. The prior work experience of many political appointees seldom has prepared them for the depth and complexity of a large public organization, whose recent history and standard operating procedures they must comprehend before they can hope to influence the agencys' operations.[4] Compounding the difficulty of this task is the relatively short tenure, averaging something less than 24 months, that most political appointees have in office, hardly enough time for an individual to develop an effective understanding for the strategic environment, develop an agenda of desired policy changes, see those changes through a lengthy legislative or regulation writing process, and guide their implementation by the agency.[5] As noted in one study of federal political executives, "the single most obvious characteristic of Washington's political appointees is their transience."[6] Typically, this transience also aggravates another standard problem for political appointees: the lack of teamwork, and of clearly articulated shared goals and means, between the layers of political executives within an individual department and across departmental lines.[7]

Nevertheless, there is a substantial professional cadre within the office of the secretary that remains reasonably constant from one administration to the next and exhibits all the essential characteristics of an organization. A variety of complex tasks are performed in accordance with a well-developed set of standard operating procedures that largely define what the outputs will be. Predominant among these tasks is the processing of the annual budget requests from the various agencies within the Department of Agriculture: assembling the budget requests, presenting an analysis of budget options for each agency to the responsible political official, and recasting the budget request to reflect the decisions of the secretary of agriculture in the Department Allowance that is submitted to OMB. Being essentially a staff organization, the concerns of the professional cadre within the office of the secretary are based more on the process, rather than the particular substance, of the decisions being made. Budget analysts in the secretary's office seek to explode agency budget requests into their component parts, reexamine the assumptions underlying the amounts requested for particular activities and discover any additional decision options that were not displayed by the agency itself in the original budget request. As this independent analysis is being performed, the analysts are continually mindful of the policies, priorities and preferences of the secretary and particularly of the assistant secretary who holds direct line authority over the agency whose budget is under review. Efforts are made to illuminate aspects of the budget that relate directly to those concerns. The analysis is also necessarily a reflection of the particular concerns and expertise of the analyst. The number of budget analysts in the secretary's office is relatively small, with the responsibility for the review of a particular agency's budget often resting largely with one individual. The tenure of those individuals tends to be relatively long, usually spanning several presidential administrations. The filter through which the agency request must pass to reach the political decision makers is thus long and often has but a few passages; whether intentional or not, the output will reflect the inclinations, interests and accumulated experience of the analyst. In the sense of process, if not substance, the options presented to the decision maker are defined and limited as an organizational output of the professional staff within the secretary's office, just as the original proposal is an organizational output of the agency.

The Office of the Secretary of Agriculture has gone through a series of organizational changes in recent years that are of relevance to how the Forest Service budget is dealt with. Below the secretary and deputy secretary are a number of assistant secretaries, each with responsibility for one or more of the agencies in the department. The exception to this is the assistant secretary for administration, whose has responsibility for the routine tasks of the secretary's office itself, including the collection and processing of the agencys' budget requests. Prior to 1975, policy matters for the Forest Service were dealt with by the assistant secretary for conservation, science and education, whose responsibilities also extended to the Soil Conservation Service, the Agricultural Research Service, the Cooperative State Research Service and the National Agricultural Library. In 1975, the research responsibilities were transferred to a newly created office of the director of science and education, leaving the renamed office of the assistant secretary for natural resources and environment with responsibility only for overseeing the operations of the Forest Service and Soil Conservation Service. Partially due to this narrowing of responsibilities and in part perhaps to coincidence, the individuals serving as assistant secretary for natural resources and environment after this time have assumed a much greater personal involvement in policymaking, and even the day to day operations, of the Forest Service than ever before.

As a part of this reorganization of the office of the secretary in 1975, the assistant secretary for administration sought a greater policy role in dealing with agency budgets, rather than simply collecting and assembling the paperwork. The Office of Budget and Finance was established and included economists and policy analysts whose role was to look at longer term implications of particular policies and how policy changes could be accomplished using the department budget allowance as the vehicle. The assistant secretary for administration very soon found himself tangling with each of the assistant secretaries with direct agency responsibilities. In an effort to avoid such perennial and institutionalized squabbling, then-Deputy Secretary Richard Lyng in 1981 removed the policy analysis responsibility from administration and established it as an independent office, the Office of Budget and Program Analysis (OBPA), reporting not to any assistant secretary but directly to the deputy secretary. Nonetheless, OBPA continues to provide staff assistance to the assistant secretaries as well as the deputy secretary in the analysis of annual budgets from the agencies and longer term analyses of particular policy issues and initiatives. Again, the individuals involved are few and tenure is long; the personal interests and inclinations of the analysts tend to determine what analyses are done and how the policy options are defined for the political executives. In the case of the Forest Service, there is at least one policy analyst in OBPA who decades ago took a particular interest in the agency's activities and has produced "an unending stream" of analyses that, with considerable accumulated expertise, have provided the secretary's office with a valuable independent view of the Forest Service policies.

Since its very establishment in 1905, the Forest Service has enjoyed an extraordinary degree of independence from the political leadership in the Department of Agriculture. Historians have attributed great political shrewdness to the Forest Service's founding leader, Gifford Pinchot, in his maneuvering the Forest Service, and the national forests, into the Department of Agriculture rather than the Department of the Interior.[8] Within the Department of Agriculture, where the political leadership is largely preoccupied with major issues of agricultural policy, the Forest Service has been able to pursue its own goals with much greater freedom from political interference than would likely have been the case had it been placed in the Department of the Interior, whose responsibilities revolve around the management of federal public lands. Also, as a matter of general principle, Pinchot favored the "responsible exploitation" of federal forest resources "much like a crop," and saw the Agriculture Department as more amenable to this viewpoint than the Interior

Department, where officials at the time "tended to side with wilderness preservationists and to favor creation of parks and recreation areas in the forest reserves."[9] In 1897, President Grover Cleveland had sequestered 109 million acres of western public domain lands for forest reserves and placed them under the Department of the Interior. In 1905, Pinchot used this preservationist characterization of the Interior Department to parlay the continuing outrage of western development interests in Congress over Cleveland's action into the transfer of the forest reserves, later called national forests, to the Department of Agriculture and the Forest Service.

Much of the Forest Service's independence within the Department of Agriculture also stems from its tradition of being able to maintain a strong independent power base in the Congress. Again, founding leader Pinchot had provided an exemplary model in his cultivation of not only congressional support for the agency but the direct sponsorship of the President. As a condition for accepting the position as head of the Agriculture Department's Forestry Division in 1898, Pinchot extracted assurances from Secretary James Wilson, recorded by Pinchot in his diary:

> If I would take the Forestry Division I could run it to suit myself. I could appoint my own assistants, do what kind of work I chose, and not fear any interference from him.[10]

The promise was kept and, as one historian notes, "during the next twelve years, as the two men served under three administrations, the cabinet superior gave his hyperactive forester almost totally free rein."[11]

Today, the Forest Service's independent power base in Congress rests less on the personal charisma of the chief than on the demographics of the Forest Service as an organization and their overlap with the fundamental political interests of members of Congress. As a highly decentralized organization, the Forest Service has offices in well over one-third of the 435 congressional districts across the nation, and nearly *every* congressional district that is not within an urban metropolitan area.[12] Forest supervisors and regional foresters are encouraged to maintain active relationships with the senators and congressmen for the areas within their jurisdiction. The agency's demographics have given it a variation on the traditional "iron triangle" concept, with the leg of the triangle linking the agency and its key Congressional committees being exceptionally strong (and the leg linking the agency and its interest groups perhaps less developed than that of many other federal agencies). With so much of the operational authority in the Forest Service vested in the regional foresters, it can be argued that the predominant role of the chief and of the Washington office generally is to foster and maintain good relations with Congress, especially with regards to decisions on the agency's budget. This has afforded the Forest Service an independent budget channel around the secretary's office and, to some extent, around OMB. In the words of one budget analyst in the secretary's office, "at budget time, the chief wears out the pavement on Independence Avenue," the Washington, D.C. artery running between the Agriculture Building and Capitol Hill.

The Office of Management and Budget

The Office of Management and Budget (OMB), located in the Executive Office of the President, plays a central role in the functioning of federal agencies through its fiscal and legislative clearance, program coordination and budget preparation. Although there is a general impression of OMB as the implacable fiscal conscience of the government, the "abominable no-man" as it has been called, OMB as an institution remains an enigma even to many insiders in federal policymaking. The cloak that seems to cover OMB is largely of its own making; the agency does not invite scrutiny from the outside world and, as noted above, it has taken effective measures to avoid having the substance of its budget and policy negotiations with the agencies and departments widely known. Given the apparent significance of its influence on the Forest Service budget, it is important to gain a better understanding of the fundamental goals and values, and thereby the organizational outputs, of OMB.

The OMB we see today, even more than most organizations, is a function of its history. The predecessor of OMB, the former Bureau of the Budget (BOB), came into being in response to the need for debt management and the coordination of agency budget requests through the executive office. Prior to 1921, agency budgets were submitted directly to the Department of the Treasury and aggregated with little modification for use by Congress. The president played a very limited role in the formulation of a national budget.[13] In 1912, President Taft's Commission on Economy and Efficiency recommended that a national executive budget be developed "whereby the executive may be made responsible for getting before the country a definite, well-considered, comprehensive program."[14] The Congress was well aware of a recent string of budget deficits (a new phenomenon at that time) and, perhaps recognizing its own inability or unwillingness to deal effectively with such allocation decisions, permitted much of its near-total constitutional authority over budget development to be gradually assumed by the president.[15] The Budget and Accounting Act of 1921 greatly reduced the independent influence federal agencies had had on the budget decisions of Congress by specifically empowering the new Bureau of the Budget to "assemble, correlate, revise, reduce or increase the estimates of the several departments or establishments" and to make detailed administrative studies to " secure greater economy and efficiency in the conduct of public service."[16]

Just as the fundamental values of today's Forest Service were largely shaped by the agency's founding leader, Gifford Pinchot, the basic values at OMB remain much the same as when they were set out by that agency's first Director, Charles G. Dawes. Dawes asserted that the budget bureau should maintain the highest standards of professionalism and remain impartial, nonpolitical and nonpartisan, a view subsequently referred to as the doctrine of "neutral competence." Dawes saw a strict if naive dichotomy between policymaking and administration--"only a non-political staff could do a good managerial job for a political chief executive, and . . . the best way to let the technicians make their useful professional contribution [is] to keep them thoroughly subordinated to political authority."[17] The BOB professional staff would be concerned simply with economy and efficiency in routine government business, "workers in the stoke-hole who had nothing to do with the steering of the ship."[18] Dawes also stressed that the Bureau of the Budget (BOB) must serve first as a staff agency to the president, not the various departments and bureaus in the executive branch--"the effectiveness of the budget machinery depends upon its independence of departments and its complete dependence upon the President."[19] Similarly, BOB loyalty was to the *office* of the president, not to the partisan efforts of any current or preceding administration. As such, the budget bureau viewed themselves as the elite of the civil service, the American equivalent to the respected British civil service cadre at Whitehall. It was said that if a Martian army marched on the capital, everyone in Washington would flee to the hills except the budget bureau staff, which would stay behind and prepare for an orderly transition in government.[20]

However, during the Nixon administration, the professional image of the bureau was badly tarnished by politicization. The agency had always walked a tightrope between overextending its responsibilities into the political and operating levels, and overly narrowing its role and viewpoint through concentration on routine details. Either extreme would severely damage both its professional credibility and its usefulness to the president. The budget director might at times become the presidential point-man on issues, but the bureau itself could not be seen as a source of partisan support. In 1971, the former BOB was replaced by the Office of Management and Budget and the newly created Domestic Policy Council. In Nixon's words, "the creation of the OMB represents far more than a mere change of name for the Bureau of the Budget. It represents a basic change in concept and emphasis, reflecting the broader management needs of the Office of the President."[21] Under Nixon, OMB became known as the Office of Meddling and Bumbling for its widespread interference in the internal management processes of the agencies and departments. The report of President Ford's transition team (1974) found that OMB was "too involved in departmental processes and limiting the departments' ability to come up with innovative ideas." OMB had become "an advocate of policy rather than a politically neutral analytic tool . . . imposing budgetary decisions on other federal departments and agencies that were based on political considerations rather than the economic program approved by Congress."[22]

Congress' reassertion of its constitutional authority over budgeting through the 1974 Congressional Budget and Impoundment Control Act was largely a response to such widespread abuses by the Nixon administration. James T. Lynn, President Ford's budget director and the first requiring Senate confirmation, took pains to remove the taint of politicization from OMB, symbolically moving his office from the White House back to the old Executive Office Building where it had been prior to 1971. It was in this context, and at this difficult juncture in OMB's history, that the RPA legislation came up for consideration. However, OMB's resistance to the RPA legislation, and its continuing difficulty with the implementation of RPA, are rooted not in the politicization of the institution by one administration or another, but in the enduring values and perspectives of OMB as an organization and an impartial staff agency to the president.

The OMB budget examiner assigned to oversee the activities of a given agency must act, in turn, as both inquisitor and advocate. True to the intent of the 1921 Budget and Accounting Act, the president (and thus OMB) takes as comprehensive a view of federal spending priorities as any player in any branch of the government. In serving the president, OMB must balance priorities among all the federal agencies, from national defense and foreign aid programs to highways and social welfare programs. All programs, including forest resource management, must take their places in the grand sweep of federal responsibilities. As more *de facto* budget authority has come to reside with the president, OMB has played an increasingly critical role in balancing federal spending priorities while keeping the grand total within accepted deficit limits. With the establishment of the Congressional Budget Office and the budget committees in each house by the 1974 budget reform act, Congress now has at least some institution for viewing the federal budget as a whole. Though the budget committees provide guidelines to the myriad appropriations subcommittees (organized generally along the lines of the agencies they oversee), there is still no institution that can offer the type of detailed investigation, analysis and recommendations long provided by OMB.

In conducting the preliminary analysis of an agency's budget request, the budget examiner must represent OMB's fundamental perspectives and values (i.e., economy and efficiency) in pressing for the justification of continuing programs as well as of new or expanding programs. In the role of the inquisitor, the examiner must press such questions as: What is the real public demand for this good or service? Is there a legislative or executive mandate that the federal government attempt to meet this demand rather than leaving it to state/local governments or the private sector? Is the proposed program and funding level the most cost-effective means by which the stated objectives can be achieved? The first instinct of the budget examiner is to look for an opportunity to *not* approve a new initiative. However, OMB realizes that, in the words of one budget examiner, "there are many things that need to be done, and we want to do the right thing."[23] But it is the responsibility of the examiner to press hard and be certain that any new initiative *is* well justified and that adequate analysis has been done to document and support that justification.

These justifications, and supporting analysis, become critically important when the examiner must present the agency's budget to the division chief at OMB, and when it in turn must be presented to the executive associate director for budget and to the politically-appointed deputy director and director before being made a part of the President's Budget. Thus, it is in the agency's best interest for the budget examiner to have the best possible supporting analysis for its budget proposal. The examiner is the agency's only *de facto* advocate in OMB's internal process determining what will constitute the President's Budget. If an adequate justification is not at hand, the budget examiner will seldom advocate funding a program or, if he or she does push it, the program will likely be turned down at some point along the gauntlet that must be run. Also, an examiner must be careful that his or her concerns for an agency's interests do not appear to supersede loyalty to OMB; examiners may be periodically rotated to overseeing different agencies to avoid such "going native."

OMB has been integrally involved in the development of comprehensive program planning by the Forest Service, both in the content of the enabling legislation and the development of the plans themselves, and this history has a significant bearing on how OMB now views Forest Service annual budget requests stemming, at least in part, from those planning efforts. In theory, Forest Service program plans, such as the RPA Assessment and Program and later the NFMA national forest plans, provide just the sort of information OMB needs from the Forest Service in order to justify its programs. Earlier national-level program documents (e.g., the Resource Development Program [1961] and the Environmental Program for the Future [1974]) were seen as efforts to raise the visibility--and funding prospects--of Forest Service programs, primarily with the congressional appropriations committees. These were regarded by OMB as no more than program promotion devices lacking any real analytical justification for the increased budgets that were called for. When the concept of RPA was being developed, it appeared to be just what OMB was looking for from the Forest Service. The RPA Assessment document, to be prepared on a ten-year cycle, would provide a comprehensive examination of the national renewable resource demand and supply situations and the Forest Service's production possibilities and resource capabilities. Such a document would presumably provide just the detailed analytical basis needed to justify Forest Service funding requests.

However, RPA also required the preparation of a Program document that would respond to the needs and opportunities identified in the Assessment by establishing a set of 50-year resource output goals and setting more detailed resource objectives and funding targets for the five years covered by each new Program. In submitting each new five-year Program to the Congress, the President is required by RPA to accompany it with a Statement of Policy, which essentially represents the president's commitment to implementing the new Program. With each subsequent annual budget proposal, the president must indicate where the proposed budget deviates from the budget targets in the RPA Program and the reasons justifying the deviations. Finally, RPA requires the Forest Service to prepare an annual report, which must include a comparison of the actual funding and resource management accomplishments for the year with the budget targets and resource output objectives for that year as indicated in the Program.[24]

OMB's primary responsibility is to the president, and his need for flexibility in responding to shifting budget needs and priorities. OMB disliked even the implicit constraints on the president for setting Forest Service funding at whatever level he deemed necessary given the other federal responsibilities at the moment. In particular, OMB did not like the presidential accountability to Congress implied in the requirement for a Statement of Policy laying out the president's commitment to funding the Program over its five year lifetime. Also troublesome was the potential for presidential embarrassment in the requirement that each annual budget show deviations from the Program and the reasons therefor. Similarly, any significant deviations from the Program that appeared in the annual report could be portrayed as the president's lack of commitment to conservation or his inability to carry out his own natural resource policy and programs.

In congressional testimony, OMB opposed these features of the RPA bill but, interestingly, did not oppose the underlying purpose and thrust of the legislation. The reporting requirements remained intact in the final bill, however, and OMB recommended a presidential veto. As the RPA bill reached the White House, Richard Nixon had just resigned the presidency, so the bill was among the first pieces of legislation reviewed by President Ford when he assumed the office. Seeing the strong bipartisan support for the bill in both houses, and not wanting to get off to a contentious start with the Congress, President Ford signed the RPA despite the veto recommendation.

But OMB authored the presidential caveat that went along with the signing: "certain provisions of this act disturb me, especially those provisions relating to presidential discretion in formulating annual budget requests for our nation's forestry programs."[25]

OMB's concern for protecting the president's flexibility regarding Forest Service budgets has not abated over more than a decade of RPA implementation. This concern clearly extends to the mix of funding among resource programs as well as the size of the overall Forest Service budget. With the Forest Service's issuance of the first RPA Program in 1975, the Congress took seriously the role of the RPA in establishing budget goals for national forest management activities. The sum total of the FY 1978 Forest Service request for national forest programs was higher than OMB would permit, and the proposed expenditures for several programs were scaled back in the President's Budget. Rather than using the funding distribution in the President's Budget, however, Congress simply took the budget goals for each NFS resource program displayed in the 1975 Program and superimposed a 15 percent across-the-board reduction to bring the overall budget within spending limits. It was largely OMB's consternation over this interpretation of the RPA Program budget goals that motivated the development of the "low-bound" alternative, complete with its own set of budget goals, for the 1980 RPA Program.

This low-bound alternative did not emerge without its own difficulties, however. The Forest Service was already smarting from the secretary of agriculture's modification of the budget goals associated with the original proposed Program, in which the secretary had specified that program budget decisions be based on increments of management effort on a resource-by-resource basis; the Forest Service had argued for a small number of integrated multiple-use alternatives, fearing that a separate treatment of resources might result in an overall program that was physically, economically or environmentally not implementable on the ground.[26] The protracted disagreement between the Forest Service and OMB over what amounted to a further major modification in the Forest Service's work delayed the submission of the Program to Congress by several months. A congressional resolution demanding the overdue Program finally broke the logjam, but OMB insisted on its view being presented, if not as the only Program proposal, then at least as a fully developed alternative to the proposal from the Forest Service and secretary of agriculture, now known as the "high-bound" alternative. As noted above, the low-bound alternative was rejected by Congress, though congressional appropriations to the Forest Service subsequently matched the low-bound alternative much more closely than they did the high-bound. In the development of the 1985 RPA Program, the document was again held up by OMB while negotiations took place with the Forest Service and the secretary's office. Nearly two years past the deadline, the 1985 Program was submitted to Congress with both a high-bound and low-bound alternative. Now, nearly two years later, no statement has yet been issued by any congressional committee rejecting or accepting either of the alternatives.

As an organization, OMB has exercised a great deal of control over the budget for the Forest Service and has been able to shape the budget according to its own values, even when it meant substantially departing from the goals and objectives stated in the agency's program plans. Gradually, with the passing of each new RPA Program, OMB appears to be successfully gaining a controlling influence over the Forest Service's centralized planning processes as well.

Congress

The influence of OMB over budgets for Forest Service programs pales in significance before that of the fundamental shift in budgetary power that has taken place from the Congress to the Executive Office of the President, and thus to OMB. Article I, Section 9 of the U.S. Constitution granted almost total power for determining federal expenditures to the Congress, not the executive, and up to the beginning of the twentieth century, Congress dominated the American government and its budget procedures.[27] Since that time, there has been a steady delegation of power over the budget to the president, resulting in the budget being initiated and to a great extent dominated by the president; "[t]his has ordinarily not been the result of conspiracy or usurpation but of deliberate delegation, atrophy, or shunning of power by Congress."[28] With the exception of a brief period during the Nixon administration, when Congress sought to recover through the 1974 Congressional Budget and Impoundment Control Act some of the budget power it had let fall to the president, there has been a steady flow of authority from the legislative to the executive branch. This delegation of power has been quite deliberate, "the product of considered legislative decision, neither acts of impulse by the Congress nor presidential *coups d'etat* . . . [t]he Congress has to consent, because it has to pass the laws."[29]

The law that has served as perhaps the historical watershed in the relationship of Congress and the president regarding budgeting was the 1921 Budget and Accounting Act that established the Bureau of the Budget. Born of Congress' inability or lack of political will to keep spending within limits, this law gave the BOB central authority to coordinate and control the budget requests of all federal agencies through the submission to Congress of a single President's Budget. Despite the continued existence of "iron triangle" relationships between agencies, their congressional committees and interest groups, this represented a substantial narrowing of the thoroughfare that had run between the agencies and Congress. BOB was subsequently given authority to clear all legislation proposed to Congress. In 1933, BOB was given authority not only over budget preparation but budget execution as well when it gained "the power to make, waive and modify apportionments of appropriations, a responsibility previously fixed in the individual department heads and bureau chiefs."[30] The Anti-Deficiency Act, passed in 1950, uses the threat of fines and jail to prevent agency leaders from increasing their appropriations by spending all their funds during the first three quarters and then going to Congress for "emergency funding" to keep the agency from having to close down (the so-called Washington Monument ploy).

In a reorganization plan approved by Congress in 1970, BOB became OMB, its scope expanded such that "[i]ts authority to act in the president's name extends to the entire Federal administrative establishment . . . it keeps its finger on every pulse of Federal activity."[31] The authority to "act in the president's name" is no small part of the power wielded by OMB. The symbolism extends to geographic location as well as activity. The 1939 transfer of BOB from the Treasury Department to the Executive Office of the President placed the BOB director next door to the White House; Nixon's 1970 reorganization put the director *in* the White House, only to have it symbolically moved out again following the collapse of the Nixon presidency.

But the symbolism that is most important to the president's power over the budget is the high ceremony and almost religious significance accorded the President's Budget (always capitalized) as it is released to Congress. This ritual is carefully orchestrated by the White House and OMB's political leadership to enhance the president's power in two ways: (1) keeping the agency troops in line, and invoking the name of the Office of the President to make what agency leaks there are seem petty and pedestrian; and (2) presenting an organized and unified front to force the Congress to react, and to capitalize on the contrasting disarray. The release of the President's Budget is one of the most awaited and publicized events on the Washington calendar.

> When the budget comes down to Congress, it is as though it were Holy Writ newly discovered. It is wrapped in the gold paper of the Executive Office of the President; it is tied and bound and finally crowned with the sealing wax of the presidential seal. It is opened and revealed to the waiting publicans with an almost religious mystique.[32]

The proposed budget affects the interests of each member of Congress differently, depending on which region of the country he or she represents or what interests he or she tends to act as advocate for. Shortly after the budget becomes publicly available,

> the reaction is swift, hearty, and partisan. There are few

> introverts in the House or Senate, and members . . . neither hesitate to react nor hide their opinions.[33]

Budget testimony by the budget director, other cabinet officers, and agency leaders, having already been regimented and coordinated through the largely invisible internal negotiations of the past months, seems calm, coherent and reasoned by comparison to the chaos of congressional reaction.

> The momentum lies with the president and his "team," who speak in harmonious tones against the discord and babble of several hundred disparate voices on the Hill. . . . A forceful and determined president backed by his cabinet spokesmen and defended by his congressional partisans can overwhelm Congress in its confused, disordered and chaotic state.[34]

Not by chance, the public image of the Congress, and thus to a considerable extent its power relative to the president's, is not enhanced by this annual spectacle. Observed Hubert Humphrey, former congressman, senator and vice-president:

> Congress is on the defensive, able merely to attack, adjust, minimize, expand, or contract what has been presented by the president. But the *initiative* for planning, for determining the amount and timing of expenditures, is in the hands of the president. Congress, no matter how well run, how wisely populated, seems to haggle, delay, argue, debate while the country and the president wait.[35] [emphasis in original]

Yet despite the way the current system allows the president to put the Congress off balance, and seemingly gives away what would appear to be a major strategic advantage for Congress provided by the Constitution itself, it is unclear that the Congress could function otherwise to provide a federal budget each year. Congress finds it difficult to do its job without presidential leadership, without the "great initiator" providing something for them to react to. Neustadt asserts that Congress relies heavily on this role of the president as an "invaluable clerk," that

> [s]ervice for themselves, not power for the president, has brought them to accept his leadership in form . . . Congressmen need an agenda from outside, something with high status to respond to or react against. What provides it better than the program of the president?[36]

Against this backdrop, the House begins its formal hearings and deliberations on the budget for the Forest Service. By convention, all appropriations bills are initiated in the House and are taken up sequentially by the Senate, with chairmanship of the joint conference committee alternating between the two on an annual basis.[37] In viewing Congress in organizational terms, the most structurally useful approach is to look at the committees and subcommittees rather than at Congress as a whole. There are no fewer than twenty committees and subcommittees that have some sort of jurisdiction over Forest Service activities. However, the vast majority of congressional activities related to the Forest Service take place in the appropriations committees and two standing committees in each chamber: in the House, the Committee on Interior and Insular Affairs and Committee on Agriculture, and in the Senate, the Committee on Energy and Natural Resources and the Committee on Agriculture. In each of these committees, the action relating to the Forest Service is further dominated by the subcommittees with particular jurisdiction over the agency. In the appropriations committee, the Forest Service budget is not dealt with by the subcommittee that funds most of the programs of the Department of Agriculture, but is lumped together with most of the programs of the Department of the Interior and dealt with by the appropriations subcommittee for "Interior and Related Agencies."[38] The Senate subcommittees, and their jurisdictions, are mirror images of those in the House.

There are a few differences between House and Senate rules that are of direct relevance to how the Congress deals with the Forest Service budget. Chairmanships of committees and subcommittees are determined in both chambers largely on the basis of seniority.[39] In the Senate, chairmanships of subcommittees are determined on the basis of seniority in the full committee. In the House, however, seniority is determined in the subcommittees themselves, resulting in a number of situations in which subcommittees are chaired by members who are still several rungs down in seniority on the full committee. A subcommittee chairmanship is key in the final stages of committee action, the "chairman's mark," when the chairman and ranking minority member become disproportionately influential in the outcome of the bill. The other important difference is that, in the House, members usually serve on only one major committee; in the Senate, with less than one-quarter the membership of the House, senators serve on several committees and may chair more than one. This provides a different sort of advantage in each chamber. In the Senate, it is possible for an individual senator to assume a position of extraordinary influence over a particular agency by chairing both the standing committee that deals with legislation pertaining to that agency, and of the appropriations subcommittee that controls its funding. This was the case for the Forest Service from 1981 to 1986 when Idaho Senator James A. McClure chaired both the full Energy and Natural

Resources Committee and the Interior appropriations subcommittee that controls Forest Service funding. On the other hand, any given senator's attention must be diffused over a greater span of responsibilities. The advantage this gives the House is that members can concentrate their efforts on becoming much more knowledgable on the details of the particular agencies and programs within their jurisdiction, a factor of considerable importance in the final House and Senate conference on the appropriations bill.

Even in the House, however, it is by necessity the appropriations subcommittees that have the greatest influence on an agency's funding, and the chairman and ranking minority member, and their staffs, who have the greatest influence on the subcommittee's actions. Subcommittee staff members, whose tenure and expertise may exceed that of their counterparts at OMB or even the Forest Service itself, analyze and evaluate the details of the President's Budget for the agency (and often inside information on how it differs from the original agency request). Key staff members work closely with the chairman and ranking minority member at every step of the process, from questioning the executive branch witnesses during the hearings to deciding strategy in the committee leadership's private sessions.[40] The subcommittee may hold several days of hearings on the budget for a particular agency, listening to testimony first from the political executives and agency heads and then from groups and organizations whose interests will be directly affected by the agency's appropriations. Following this, the subcommittee will "mark up" the bill, going over it line by line and making additions or reductions to individual budget line items, and composing the language of the committee report, which can go into much greater detail than the bill itself and is binding upon the agency. For instance, the appropriations bill may specify a particular amount for recreation construction nationwide; the committee report may further specify the exact portion of this appropriation to be spent on the construction of a particular recreation facility on one national forest in a certain congressional district. In the words of one Forest Service budget official, "we read the bill, of course, but the committee report is where we get our marching orders."

Between the subcommittee hearings and the mark-up, both of which are open to the public, there is a private stage in which most of the critical decisions are in fact made. It has been argued that many of the devices designed to enhance the appropriations committees' ability and will to withstand spending pressures have been severely weakened over the past two decades. These include the selection of committee members from "safe" districts where reelection prospects were favorable (and members could thus afford to withstand pressures from interest groups) and the conduct of subcommittee business in closed "executive sessions" that did not provide ready access to interest groups.[41] While committee and subcommittee meetings are now generally open to the public, much of the substantive negotiation still takes place behind closed doors. The "chairman's mark" is a private meeting between the subcommittee chairman and ranking minority member and their staffs in which the overall level of the appropriation will be decided upon and funding for individual programs adjusted according to the preferences and priorities of the two leaders; "with rare exceptions, 95 percent or more of the items and the funds are decided in this unknown, unpublicized, secret conclave."[42] Only those few issues the two leaders cannot agree upon are left to be taken up by the subcommittee itself. This is also where the two leaders combine forces to plot their strategy for getting their favorite programs, now buried deep in the individual line items of the budget, through the Senate. Funding for certain programs is added as "trading material" to be given up during the conference with the Senate; or an item known to be dear to a particular senator will be deleted so that it can be added back in conference--along with some item important to the House that is likely to be deleted by the Senate, of course. The language of the committee report is also drafted in these sessions, to be presented to the other subcommittee members at the start of the subcommittee mark-up. The decisions seemingly made at the subcommittee mark-up have, in fact, largely been made beforehand; the mark-up serves as the opportunity to officially disclose these decisions to the rest of Congress and the public.

This is not to say that other members of the subcommittee, or of Congress more generally, have no impact on decisions regarding an agency's funding. The primary legislative purpose of the appropriations committees is to function as guardians against the spending interests not only of the executive branch but of Congress itself.[43] However,

> the main constraint on the appropriations committees has been their awareness that they must satisfy the expectations of their parent bodies. No legislative committee can ignore the desires of the chamber in which it resides. . . . Over the past century, Congress has bolstered the appropriations committees when it has been willing to abide by their discipline; it has weakened them when it has preferred a freer rein to spend as it wills.[44]

An informal rule exists among appropriations committee members, and among the members in both chambers generally, that funding proposals by a member, and of direct benefit to his or her district, will usually not be blocked by another unless the proposal is also clearly detrimental to some other member's district. This hallowed principle of "reciprocity," and the "pork barrel" behavior it engenders, have been stemmed somewhat, or at least driven further underground, by the establishment of the congressional budget committees and budget reconciliation requirements of the 1974 Congressional Budget and Impoundment Control Act, and further by the Damoclean threats of the 1984 Balanced Budget and Emergency Deficit Control Act ("Gramm-Rudman-Hollings"). For many years, the appropriations committees in both chambers have been dominated by southerners, but the Interior appropriations subcommittees have always had substantial representation from western interests as well, owing to the proportion of public lands in the West and the importance of their resources to economic development in the region. Recognizing the central role of this subcommittee in determining the volume of timber that can be sold from the national forests, states where the forest products industry comprises a significant segment of the economy, such as Oregon, Washington and Idaho, have maintained strong representation in spite of turnovers in the state delegation. Whether or not members from such states have actually occupied leadership positions on the subcommittee, the subcommittee chairman's recognition of the importance of this particular program to these members has, in recent years, resulted so consistently in increases over the agency request that this can almost be regarded as a standard operating procedure of these subcommittees as organizations. However, this deference by the chairman is, again, contingent upon the needs of these other members not impinging greatly upon programs of high priority with the chairman and ranking minority member. In the House at least, this situation in regards to the national forest timber sale program began to change fundamentally in FY 1986, in effect destabilizing the strategic relationship between players described by Wildavsky and forcing a reassessment of the roles played by each organization.

The power of the subcommittee leadership is moderated somewhat by the right of other members to offer amendments during the subcommittee mark-up, later in the full appropriations committee mark-up and finally on the floor of the House. Once there is an agreement between the subcommittee chairman and the ranking minority member during the chairman's mark, however, they close ranks to protect their product from amendments that might weaken the bill or threaten the delicate balance of interests it represents. Proxies are gathered from absent members of both parties to aid the subcommittee leaders in voting down any amendments offered during the subcommittee mark-up. The process is repeated in the full committee mark-up and the subcommittee leaders are the "floor managers" of the bill when it reaches the full House for a final vote. The likelihood of an amendment to an appropriations bill carrying in spite of the wishes of the subcommittee leaders diminishes with each step and becomes quite remote once the bill reaches the floor. Just as Fenno found in his review of appropriations decisions during the 1960s, few recommendations of the House Appropriations Committee are overturned on the floor.[45] Though not as invincible as they were thirty years ago, the appropriations committees, and in particular the subcommittee leaders, still wield a great deal of the power in Congress. To understand the tendencies and organizational outputs of the subcommittees is to largely understand the likely outputs of the Congress.

At this point in the process, the treatment of the original budget request has become highly fragmented, and its treatment in the Senate results in further fragmentation. Despite the specialization among House members and the line-by-line processing of the budget request, the House's attention is often focused on changes to a relatively small number of budget line items, leaving the majority of the budget request virtually unchanged.[46] The Senate appropriations subcommittees pursue an independent agenda but still act to mitigate changes made in the House, further fragmenting the original proposals by focusing primarily on a narrow slice of the overall budget. The process in the Senate is dominated by the appropriations subcommittee chairman and ranking minority member, much as in the House, through the mark-ups and on the floor of the Senate. Strategy is directed first to withstanding substantive amendments by other members of the Senate, then to the establishment of a position from which to negotiate with the House in the conference committee. In the case of the Interior appropriations bill for FY 1986, threats of weakening amendments by other members and the likelihood of a successful amendment by the chairman of the Budget Committee to substantially reduce the funding in the bill across the board moved the floor managers of the appropriations bill to withdraw it until enough favorable votes could be lined up in advance. This never occurred, and the Interior appropriations bill was never brought to a vote by the full Senate.

The structural factors that give the House an early advantage in the appropriations process, their greater expertise through specialization and their ability to concentrate their efforts on much fewer major responsibilities than their Senate counterparts, come into full play during the final conference.[47] In many areas of federal legislation, especially appropriations, the Senate is thought to dominate the conference committees in terms of having more of its provisions adopted than those of the House. However, this does not necessarily signify that the Senate thus dominates the congressional impact on public policy. Because the House acts first, the Senate is often in a position of reacting to the agenda set by the House. Though the Senate may have an advantage in terms of agreement on specifics, this may be offset by the House's ability to set much of the agenda for eventual compromise.[48] Throughout the decade 1977-1986, the correlation of the House and Senate changes to the Forest Service budget proposal with the final appropriation was virtually identical (see Table 2.5). On a program-specific basis, the Senate had a slightly higher correlation for the timber, minerals, protection and lands programs. The House had a slightly higher correlation in recreation, range, fish and wildlife, soil and water, and facilities. In fact, in the case of Forest Service budgets, probably neither side can be said to clearly dominate the outcome of the conferences. In the words of one appropriations committee staff member, "each chairman gets what he wants and can go back to his most vocal interests and show a victory."[49]

For an agency like the Forest Service, the appropriations bill itself may not be as important as the joint committee report that accompanies it and provides specific spending directives and other "marching orders." The committee report becomes an important part of the legislative history of the appropriations bill and is binding upon the agency. Yet because it is not the appropriations bill itself, it can contain many directives that would, were they contained in the bill, be subject to a point of order against legislating on an appropriations bill. The significance of this document is not lost on the conferees, of course, and there have been instances when an agreement has been reached on the language of the appropriations bill itself, only to have it collapse due to continuing disputes over the committee report language. As will be recalled, such was the case with the FY 1986 year-long continuing resolution (H.J. Res. 465) that, in the absence of an actual Interior appropriations bill, contained the funding for the Forest Service (and dozens of other agencies).

The funding of federal activities through such a massive continuing resolution instead of through the normal thirteen separate appropriations bills is a relatively new phenomenon, one regarded by many as a symbol of Congress' continuing inability to make the difficult political choices necessary to keep the federal budget within reasonable limits. Authority for continuing resolutions was provided by the 1974 Congressional Budget and Impoundment Control Act as a safety valve "for specific ongoing activities in cases where the regular fiscal year appropriation for such activities has not been enacted by the beginning of the fiscal year."[50] Few legislators envisioned that, on a regular basis, so many of the appropriations bills would fail to be reported out not only by the October 1 start of the fiscal year but by the end of the legislative session in December, and that the funding for nearly all federal activities would be rolled into one massive measure run through Congress, quite literally, at the eleventh hour. Observed former Senator Thomas Eagleton:

> This hectic, frantic, helter-skelter way of doing the nation's business is unacceptable. To rush through, on a last minute basis, the bill which funds 78 percent of appropriated monies makes a mockery of the Senate tradition as "the world's greatest deliberative parliamentary body." This year's continuing resolution stands as a monument to non-deliberation.[51]

Perhaps W. F. Willoughby, a member of the 1918 Taft Commission on Economy and Efficiency that laid the groundwork for the 1921 Budget and Accounting Act, was on to something when he recommended that "the legislature should be largely, if not wholly, excluded from the direct determination of the appropriations of funds."[52] Such a large continuing resolution is propelled by the immediate necessity of its being passed and signed by the president in order to avoid not only the halting of government activities but also the necessity of having to start the legislative process all over again in January. Confronted with this juggernaut, the hands of the president and congressional leaders are virtually tied. The resolution becomes a "Christmas tree" with hundreds of riders appended to it, yet it must be passed. This adds considerable uncertainty for agencies regarding the outcome of their budgets. At the very least, it delays their funding until at least one quarter into the new fiscal year, limiting their ability to hire personnel and set program levels that will allow for the orderly and rational expenditure of funds through the end of the fiscal year. This is only one of the many considerations that agencies must factor into the execution of their annual budgets. The particular considerations of the Forest Service in this regard will be the subject of the next section.

NOTES

1. Wildavsky, Aaron, *The Politics of the Budgetary Process* [Third Edition] (Boston: Little, Brown and Company, 1979), pp. 11-56.

2. Davis, Otto A., Dempster, M. A. and Wildavsky, Aaron, "A Theory of the Budgetary Process," *American Political Science Review*, Vol. 60, No. 3 (1966), pp. 529-547.

3. Ibid.

4. Macy, John W., Adams, Bruce, Walter, J. Jackson and Mackenzie, G. Calvin, *America's Unelected Government* (Cambridge, MA: Ballinger, 1983).

5. Brauer, Carl, "Tenure, Turnover, and Postgovernment Employment Trends of Presidential Appointees," in: Mackenzie, G. Calvin (ed.), *The In-and-Outers: Presidential Appointees and Transient Government in Washington* (Baltimore: Johns Hopkins University Press, 1987), p. 174-179. Statistics cited by Bauer are from a 1985 study of presidential appointees during the period 1966-1985 by the National Association for Public Administration.

6. Heclo, Hugh, *A Government of Strangers: Executive Politics in Washington*, p. 103.

7. Ibid., p. 104-105. See also: Corson, John J., and Shale, R. Paul, *Men Near the Top: Filling Key Posts in the Federal Service* (Baltimore: Johns Hopkins Press, 1966).

8. Cooper, John M., "Gifford Pinchot Creates a Forest Service," in: Doig, Jameson W. and Hargrove, Erwin C., (eds.), *Leadership and Innovation: A Biographical Perspective on Entrepreneurs in Government* (Baltimore: Johns Hopkins University Press, 1987).

9. Cooper, "Gifford Pinchot Creates a Forest Service," p. 72.

10. Ibid.

11. Ibid.

12. USDA Forest Service, *Land Areas of the National Forest System*, FS-383 (Washington, D.C.: USDA Forest Service, 1987).

13. Berman, L., *The Office of Management and Budget and the Presidency, 1921-1979* (Princeton, N.J.: Princeton University Press, 1979), p. 3.

14. U.S. Commission on Economy and Efficiency, *The Need for a National Budget*, House Document 854 (Washington, D.C.: Government Printing Office, 1912). See also: President's Commission on Economy and Efficiency, *Message of the President Submitting for the Consideration of Congress A Budget*, Senate Document 1113 (Washington, D.C.: Government Printing Office, 1913).

15. Shuman, H. E., *Politics and the Budget: The Struggle Between the President and the Congress* (Englewood Cliffs, N.J.: Prentice-Hall, 1985).

16. Executive Office of the President, Bureau of the Budget, *Staff Orientation Manual*, 1945, p. 40.

17. Price, Donald K., "General Dawes and Executive Staff Work," *Public Administration Review*, Vol. 10, No. 2 (1941), p. 169.

18. Ibid.

19. Dawes, Charles G., *The First Year of the Budget of the United States* (New York: Harper Brothers, 1923.), p. 63.

20. Berman, *The Office of Management and Budget and the Presidency, 1921-1979*, p. x.

21. *Weekly Compilation of Presidential Documents*, March 6, 1970 (Washington, D.C.: Government Printing Office, 1970), pp. 355-357.

22. Recommendations cited in: U.S. Senate, Committee on Governmental Affairs, *Nominations of Thomas B. Lance and James T. McIntyre, Jr.* (Washington, D.C.: Government Printing Office, 1977), p. 40.

23. Personal communication, Office of Management and Budget, Natural Resources Division, August 4, 1987.

24. See: Forest and Rangelands Renewable Resources Planning Act of August 17, 1974 (P.L. 93-378, 88 Stat. 476, as amended; 16 U.S.C. 1600-1614).

25. U.S. Senate, Committee on Agriculture, Nutrition and Forestry, *Compilation of the Forest and Rangeland Renewable Resources Planning Act of 1974* (Washington, D.C.: Government Printing Office, 1979), p. 218.

26. Fedkiw, John, "The USDA Decision Process for the 1980 RPA," in: Hewett, Charles, and Hamilton, Thomas, *Forests in Demand: Conflicts and Solutions* (Boston: Auburn House, 1982), p. 164.

27. Fenno, Richard F., *The Power of the Purse: Appropriations Politics in Congress* (Boston: Little, Brown and Co., 1966).

28. Shuman, *Politics and the Budget*, p. 18.

29. Sundquist, James L., *The Decline and Resurgence of Congress* (Washington, D.C.: The Brookings Institution, 1981), p. 35.

30. Parris, Judith, *The Office of Management and Budget: Background, Responsibilities, Recent Issues*, Congressional Research Service Report 78-158 (Washington, D.C.: Library of Congress, 1978), p. 16.

31. Ibid, p. 26.

32. Humphrey, *The Education of a Public Man*, p. 193.

33. Shuman, *Politics and the Budget*, p. 60.

34. Ibid.

35. Humphrey, *The Education of a Common Man*, p. 103.

36. Neustadt, Richard E., *Presidential Power: The Politics of Leadership* (New York: Wiley and Sons, 1976), pp. 74-75.

37. This convention is the result of an odd set of events, mentioned earlier, in which the House essentially relinquished to the Senate its authority to initiate all tax bills, specifically granted by the Constitution, in order to remain unchallenged by the Senate in its right to initiate all appropriations bills, for which it has no constitutional authority. See: Shuman, *Politics and the Budget*, p. 66-67.

38. This structure is interesting in light of the repeated attempts over the years to have the Forest Service transferred from the Department of Agriculture to Department of the Interior, which contains all of the other federal land management agencies. There have been numerous reorganization attempts since 1905, the most famous and concerted being those of Interior Secretary Harold L. Ickes in the mid-1930s. Presidential commissions on improving government efficiency as recent as 1980 have continued to recommend the transfer, and the Forest Service, with the aid of powerful supporters in Congress, has continued to strongly resist any such reorganization [see: Robinson, Glen O., *The Forest Service: A Study in Public Land Management* (Baltimore: Johns Hopkins University Press for Resources for the Future, 1975), pp. 13-14; Ickes, Harold L., *The Secret Diaries of Harold Ickes: The First Thousand Days* (New York: Simon and Schuster, 1953)]. Such a transfer, were it to eventually be made, would necessitate few changes in the current committee jurisdictions in Congress, however. Indeed, the current organization of the legislative branch would imply that the transfer has, in effect, already been made, even if the organization of the executive branch does not. Already, the vast majority of legislation affecting the Forest Service comes through the Interior and Insular Affairs Committee (Subcommittee on Public Lands and National Parks) in the House and the Energy and Natural Resources Committee (Subcommittee on Public Lands, Reserved Water and Resource Conservation) in the Senate, both of which primarily oversee the activities of the Department of the Interior. The agriculture committees in both chambers retain jurisdiction and receive joint or sequential referrals of bills but seldom initiate legislation affecting the Forest Service. The Agriculture appropriations subcommittees in both chambers have long since relinquished any jurisdiction over Forest Service appropriations, leaving this entirely to the Interior Appropriations Subcommittees.

39. In recent years, committee chairmanships have been decided in the majority party caucus that takes place prior to the start of each new Congress. However, the all-important subcommittee chairmanships are still decided within the committees themselves and usually go to the ranking majority member on each subcommittee.

40. Personal communication, staffs of Senate and House appropriations committees, U.S. Congress.

41. Schick, Allen, *Congress and Money: Budget, Spending and Taxing* (Washington, D.C.: The Urban Institute, 1980), pp. 427-428.

42. Shuman, *Politics and the Budget*, p. 72.

43. Schick, *Congress and Money*, pp. 416-418.

44. Ibid.

45. Fenno, *Power of the Purse,* pp. 448-449.

46. Fenno, Richard, "The House Appropriations Committee as a Political System: The Problem of Integration," *American Political Science Review,* Vol. 56, No. 2 (1962), pp. 310-324; Fenno, Richard, *The Power of the Purse.*

47. Shuman, *Politics and the Budget*, p. 80.

48. Ripley, Randall B., *Congress: Process and Policy* [Second Edition] (New York: W.W. Norton, 1978), pp. 181-186.

49. Personal communication, Senate appropriations committee staff.

50. Congressional Budget and Impoundment Control Act of 1974 (P.L. 93-344), p. 70.

51. Washington *Post*, December 21, 1982, p. A13, as cited in: Shuman, *Politics and the Budget*, p. 81.

52. Willoughby, W. F., *The Problem of a National Budget* (New York: Appleton, 1918), p. 39.

6

Applying the Conceptual Models to the Decisions on the Fiscal Year 1986 Forest Service Budget

The Forest Service Budget as an Output of Organizational Processes

Organizational Processes within the Executive Branch

The budget for the management of the National Forest System, as contained in the President's Budget for Fiscal Year 1986, was a real-term reduction of slightly more than ten percent from the prior year's congressional appropriation. In the first analysis, this seems to be an eminently rational response to the major issue and driving policy underlying the federal budget development process at this time: the need to reduce the burgeoning federal deficit, without resorting to additional taxes, through substantial cutbacks in all domestic federal programs except defense and entitlements. The budget justification submitted to Congress with the FY 1986 President's Budget implies in general terms that this is indeed the unifying theme in the budget proposal. In terms of the Rational Policy Model, this large-scale action is a rational response to a large-scale cause. But this explanation alone offers little information as to what other factors may have been considered specifically in relation to the management of the national forests. Certainly the major, government-wide thrust to reduce the deficit by bringing down spending, led through a high-visibility campaign by the president himself, accounts at least in part for this action. But what are the other factors, the ones *not* described in the budget justification? What were the roles of the various organizations within the executive branch that shared in developing, refining and transforming the Forest Service budget request into the specific program that eventually appeared in the President's Budget?

The budget proposed for the National Forest System in the FY 1986 President's Budget was, in large part, a consequence of the established routines and procedures of the several organizations that took part in the process. Two years before, the Forest Service had been given instructions to prepare an FY 1986 budget proposal based upon the funding shown in the FY 1984 President's Budget. The overall NFS funding in the resulting agency request matched that in the FY 1984 President's Budget almost precisely--and in real terms was actually more than a 4 percent reduction. The agency request was also within 1 percent of the prior year's (FY 1985) congressional appropriation for NFS activities, which again represented a net reduction in real terms. The proposed timber sale level of 11.2 BBF, plus an anticipated 0.5 BBF in reoffered timber sales, equaled the total timber sale level assigned by Congress in FY 1984 and matched with what the Forest Service projected would be the actual timber harvest level in FY 1986. The reoffering of timber sales that had been turned back to the agency under the terms of the Federal Timber Contract Payment Modification Act permitted the Forest Service to project the same level of timber sales even while proposing a nearly 4 percent real-term reduction in funding for the timber program. The leeway provided by the reduced timber costs allowed the Forest Service to propose generally 5 to 6 percent real-term increases for the much smaller NFS nontimber resource programs and still stay within overall constraints. Apparently, the well-established bureaucratic process of budget development, with progressive refinements at each step up the Forest Service administrative hierarchy, had produced a result very close to the mark.

A longer term perspective suggests, however, that certain fundamental aspects of the Forest Service's internal budget development process were not so well established. A major change in the budget development routine had taken place when the RPA Program budget targets lost their central role in the agency's development of budget alternatives. The RPA budget targets no longer served as the benchmark around which lower and higher alternatives were developed; beginning with the FY 1986 budget, the RPA targets themselves were relegated to being an alternative to the new central benchmark, one based on the classic incremental approach to budgeting. This change may have had a greater impact on budget development routines at the field offices of the Forest Service than on budget strategists in the Washington office. It can be argued that although the standard operating procedure since 1977 had been to use the RPA budget targets as the centerpiece for annual budget development, the agency's top-level strategists had taken a more pragmatic and flexible approach based on their most up-to-date assessment of what was possible in the current political environment. The funding increases sought by the Forest Service during the late 1970s were far too large to be considered a true incremental approach to budgeting, but neither did the agency requests reflect the balance of funding represented in the RPA budget targets (see Table 2.4).

Taking an Organizational Process Model approach and turning it inward on the Forest Service itself, there may be internal as well as external reasons why the agency request, viewed itself as an organizational action, never quite reflected the priorities represented in the RPA Programs. The sort of shift in program emphasis envisioned in the 1975 and 1980 RPA Programs would be expected to have caused struggles between the different program staffs over the reallocation of budget resources from commodity to non-commodity programs. In fact, the major real-term increases in the budgets for the non-commodity programs did begin to take place in FY 1978, but this was not until after the budget for the timber program was allowed to more than double in real terms from FY 1975 to FY 1977. Moreover, even throughout the period when the budgets for the non-commodity resources programs were growing, the budgets for these programs were still far below those called for in the RPA Programs. This was not due to modifications by other participants in the Forest Service budget process; the budget requests of the Forest Service itself were far below its own RPA targets for the non-commodity resource programs, though they remained very close for the timber program. The implications of this "glass ceiling" on funding for the non-commodity resource programs is somewhat mitigated by the fact that, under the former system of budget execution by the program staff directors and "timber czars," some portion of the increases in the timber budget may have been used to facilitate nontimber activities as well. This would not have been as significant a factor after the budget execution policy changes of 1982, however. As overall NFS funding began to be cut back in real terms the non-commodity programs assumed a disproportionate share of the reductions relative to the timber program, even in the agency request.

To anticipate the needs and concerns of their politically appointed superiors, the professional staffs at the Office of the Secretary of Agriculture, and in particular the Office of Budget and Program Analysis, had to address both the economic and political implications of the budget for the Forest Service. The two most salient issues at this time were (1) the continued soft timber market in the West, resulting in an unprecedented 40 BBF backlog of sold but uncut timber on the national forests, and (2) growing pressure on the Forest Service to sharply reduce timber sales in areas where timber receipts typically failed to recover timber sale costs. OBPA was also mindful of the general budgetary policies emanating from the White House and OMB: reduce federal spending, and give special emphasis to revenue-producing programs. The limited number of personnel at OBPA necessitated a strict prioritization of the time spent analyzing the Forest Service budget and, given their judgment that "setting the timber sales level is the single most important decision in determining Forest Service funding and staffing," their analysis focused primarily on the budget for the timber program. There were at least two alternatives that OBPA felt had not been adequately developed and presented by the Forest Service. Assistant Secretary Crowell had specifically asked OBPA to evaluate the possibility of increasing national forest timber sales, generally a net revenue producing program, by one billion board feet over the agency request of 11.2 BBF. On its own initiative, OBPA evaluated the possibility of maintaining the existing harvest level and simply redirecting timber sales to the more cost efficient regions rather than beginning from the assumption, as did the Forest Service, that the distribution of timber sales would have to remain basically unchanged from that in previous years.

The record does not indicate that OBPA made any clear recommendation to the Assistant Secretary, but the tone of the analysis seems to imply a preference for the latter alternative. According to their evaluation, to increase timber sales by 1 BBF over the agency request would entail additional costs of nearly $70 million, would require an increase of more than 2,000 staff-years over the current personnel ceiling, and would likely produce a net revenue loss, since "70 percent of the additional funding would go to the high-cost regions which produce only 43 percent of the increase in sales volume."[1] In the evaluation of the second alternative, termed by OBPA the "economic efficiency alternative," it was noted that costs would be higher by $19 million but receipts would be higher by $109 million, yielding a significant net gain to the Treasury. Personnel requirements would be 200 staff-years lower than in the agency request. There would also likely be resistance to the idea from members of Congress from the Rocky Mountain states where timber sales would be reduced to 80 percent of current levels. Of particular concern was Idaho's Senator McClure, who chaired the Interior Appropriations Subcommittee. Perhaps focusing more on the administration's general policy of giving preference to revenue producing programs and seeing the overall profitability of a 12.2 BBF sale level more clearly than the marginal loss associated with the additional 1 BBF, Assistant Secretary Crowell selected the first alternative.

The established routines and procedures at OMB are outgrowths of the organization's general drive to improve the economy and efficiency of government programs. The political executives at OMB tend to be responsive to the priorities and agenda of the president, but generally they are impartial to the details of the program proposals of individual cabinet departments and get involved in only a few highly visible policy controversies. The program level decisions at OMB are heavily influenced by decision rules based on the long-standing organizational values described earlier. A mode of analysis routinely applied to agency budget requests seeking increased funding for particular programs is marginal analysis, the comparison of the proposed incremental increase in program expenditures to the projected incremental increase in program benefits (to the degree that the benefits of the programs can be measured). The general White House policy of increasing revenue producing programs was secondary to this institutional standard operating procedure, leading the organization to draw a different conclusion about what should be done with the Forest Service budget than that drawn by the Office of the Secretary of Agriculture. It can be argued that the departments tend to share the broad role that Wildavsky attributed to the agencies, to "act as advocates of increased expenditure." Even in an era of unacceptable federal deficits and an overarching policy of spending reductions this remains a strong tendency, the office of the secretary considered only alternatives that would maintain or increase Forest Service funding, especially for the timber program. OMB, in its role as "presidential servant with a cutting bias," took the type of analysis performed by OBPA in its "economic efficiency alternative" one step further, not constraining the outcome to a level of spending equal to or higher than that proposed by the agency. Despite the fact that at the 11.2 BBF level the national forest timber program is a net revenue producer, the marginal analysis approach used by OMB revealed that this level of timber sales still included a significant volume for which the costs exceeded the returns.[2] In the judgment of OMB, therefore, the net marginal revenues of the timber program could be maximized at a timber harvest level somewhat lower than the 11.2 BBF proposed by the Forest Service, and a good deal lower than the 12.2 BBF proposed by the secretary. Furthermore, the low current demand for national forest timber, as indicated by the enormous backlog of uncut timber under contract, suggested to OMB that a continued high level of timber sales would only flood regional markets, driving down both stumpage prices and the net revenue the federal government would receive from the sale of national forest timber. The combination of these considerations led OMB to recommend an FY 1986 national forest timber sale level of no more than 10.2 BBF. This was later increased to 10.7 BBF in the President's Budget for reasons that will be discussed below.

Organizational Processes within the Congress

The analysis of the Forest Service budget by the "organizations" within the Congress is driven less by concerns for the continued solvency of the Treasury than by the political calculus of garnering a requisite share of the benefits of federal spending for local constituents. Again, the primary focus seems to be on the national forest timber harvest level and funding for the timber program, but for different reasons than at the secretary's office or OMB. Given that the timber program is budgeted at nearly $1 billion per year and represents three-quarters of the resource management spending within the National Forest System budget, the House in particular pays close attention to proposed funding increases and changes in staffing levels. However, a review of the record of meetings of the House Interior appropriations subcommittee over a period of years reveals that at least as much attention, if not more, is paid to the setting of the timber sale level as to the actual funding for the timber program, at times to the extent that the subcommittee takes the two in opposite directions.

For instance, as the political issue of "below cost timber sales," the sale of national forest timber at a net financial loss, has become visible enough to demand congressional attention, there has been increasing pressure to reduce funding for the timber program. The beginnings of this trend could be seen with the FY 1986 budget proposal when the House Interior Appropriations Subcommittee cut back funding by nearly $60 million--$20 million from the timber sales program directly and another $39 million from the construction of timber access roads. At the same time, however, the same subcommittee increased the timber sale volume from the 10.7 BBF in the President's Budget to 11.2 BBF. Although the timber sale volume then matched that in the original Forest Service request, the funding for these activities had been substantially reduced, once by OMB and again by the House. This could be interpreted as an indirect attempt by the House to force the Forest Service to take the sort of action described in the OBPA "economic efficiency alternative," that is, to maintain the same overall sale level but redirect the sales to the more cost-efficient regions. An unusually emphatic letter from the Forest Service to the House subcommittee pointed out that it would be impossible to maintain sale levels with such drastic reductions in funding, and that the sale program would have to be reduced by up to a third in following fiscal years due to the lack of road access. In recent years, the House Interior appropriations subcommittee has so regularly reduced timber funding while raising timber sale targets that the Forest Service now just as regularly prepares a feasibility report for the House, indicating the level of timber sales the agency estimates is actually possible given the funding approved by the House.[3]

Such actions by the House Interior appropriations subcommittee are quite rational when the forces driving it as an organization are examined. On one level, these actions are in keeping with Wildavsky's characterization of the House as the "guardian of the Treasury." A look back over the past decade or so shows timber expenditures to have more than doubled in real terms, yet the timber sale level is actually lower now than in the mid-1970s. The myriad tasks and limited time of the appropriations subcommittee staff discourages a highly detailed analysis of unit costs and the reasons for their increase, especially over a period of several years. If there is any "fat" in the budget, however, increasing the output targets while reducing appropriations is one technique, however blunt, for forcing it out. On a second level, however, there are structural factors based on current law and policy that make the setting of the national forest timber sale level an action representative of the most traditional and fundamental forces motivating a member of Congress--maintaining a sufficient flow of federal benefits to his or her constituents in order to remain elected.

A sequence of laws enacted over eight decades ago continues to give state and local governments a strong and direct financial interest in the volume of timber that is cut each year on the national forests. In the very beginning days of the national forests, Pinchot had capitalized on western development interests' outrage over

President Cleveland's "lock-up" of 109 million acres of public domain land in the West to have the national forests (then known as forest reserves) transferred to the more development minded Forest Service in the Department of Agriculture. Once this was accomplished, however, Pinchot fomented further outrage on the part of these interests by convincing President Theodore Roosevelt to carve out another 86 million acres of western public lands to add to the National Forest System. Congress' response was to revoke the president's authority to create new forest reserves by executive order.[4] To be sure that the western states would receive at least some benefit from this 194 million acres of now untaxable land, Congress in 1907 required that 10 percent of the Forest Service revenue from timber sales and fees be given over to the states for "the benefit of the public schools and public roads of the county or counties in which such national forest is situated."[5] This revenue share was raised to 25 percent in 1908.[6] When a substantial portion of the timber sale receipts began to go to the Forest Service not as cash but in the form of road construction services by timber purchasers, state governments prevailed upon the Congress to include the value of these services (recorded as credits given toward the purchase of national forest timber) in the revenue sharing formula as well.[7] OMB and the Reagan administration proposed in the FY 1986 President's Budget that the Forest Service be permitted to deduct its own administrative costs before calculating the 25 percent payment, that is, to pay to the states 25 percent of *net* receipts rather than of gross receipts. The proposal brought a sharply negative reaction from the key standing committees in both houses of Congress, which refused to take up any legislation to this effect.

This system of revenue sharing places the county and state governments, perhaps the single most influential constituent voices for any member of Congress, in the position of advocating as high a level of national forest timber sales as possible, with a minimum of fluctuation from one fiscal year to the next. In some heavily timbered counties in the Pacific Northwest, revenue sharing from national forest timber receipts constitutes more than 80 percent of the annual operating budget. Congressional representatives from districts or states with large areas of national forest land cannot sit idly by and permit a significant reduction in the timber harvest level if they hope to stay elected. Observers who attribute the emphasis on the timber program to a cultural predisposition within the Forest Service may not be aware that in nearly every year the timber sale target that Congress assigns to the Forest Service in the appropriations bill is significantly higher than the agency itself has proposed (see Table 6.1). Other observers attribute it to the influence of timber industry representatives on congressional appropriations. In areas where local industries are heavily dependent on national forest timber outputs, members of Congress pay attention because "this translates into employment in the affected industries and ultimately into votes."[8]

Table 6.1
Timber Harvest Level in the Proposed Budget for the Forest Service, Compared With That Assigned in the Final Appropriations Bill, FY 1977-1986

FISCAL YEAR	USFS	USDA	OMB	CONGRESS
[BILLION BOARD FEET]				
1977	*	*	10.4	10.7
1978	9.7	9.7	10.2	12.0
1979	*	11.5	11.5	12.4
1980	10.5	*	11.7	12.2
1981	11.7	11.9	11.9	12.1
1982	*	*	11.9	11.1
1983	8.5	12.3	12.3	11.3
1984	13.1	13.1	11.6	11.7
1985	12.3	12.3	11.2	11.2
1986	11.2	12.2	10.7	11.4
1987	10.9	10.0	10.0	11.2
1988	11.1	11.1	11.1	11.4
1989	11.2	11.2	11.2	11.5

* Not available.

Source: USDA Forest Service, Washington Office, Timber Management.

But the evidence suggests that the timber industry may be more of an incidental beneficiary of actions directed primarily to protecting the interests of state and local governments. A recent editorial in a Northwest newspaper asked whether it should be "federal policy to make county governments and local school officials captive allies of the timber industry," pointing out that this system forces local governments to advocate actions in the short term that may be against the regions' best interests in the longer term.[9] Situations can arise in which timber harvesting is ratcheted above what the forest managers regard as sustainable levels not because local governments want more timber cutting *per se*, but because they seek a stable stream of income from Forest Service receipt sharing.

During the committee mark-up sessions, the politics of federal benefit allocation define even more explicitly the organizational outputs of the committees and the Congress. The Forest Service budget is programmatic rather than oriented by physical locations. Unlike the National Park Service budget, which may have a specific line item for Yellowstone National Park or the Golden Gate National Recreation Area, the Forest Service budget does not lend itself to directing funds to a particular national forest in a given congressional district. However, as pointed out earlier, the committee report on the appropriations bill can be used to direct funds to specific activities, usually construction projects, to one national forest or another. Subcommittee members who have never attended a hearing will be there to represent the interests of their constituents in the mark-up session, as well as full committee members who may attend and participate in subcommittee mark-ups as *ex officio* members of the subcommittee.

The mark-ups can also focus on programmatic aspects of the appropriations bill itself, especially when the link between the actual funding for something like the timber sale level and economic activity in individual congressional districts can be easily visualized. The political football for FY 1986 was funding for the construction of timber access roads which, as pointed out by the Forest Service, would have direct impacts on future timber sale levels. This item had taken a reduction of $50 million in the subcommittee, had $11 million of this added back in full committee mark-up and fell back to the $50 million reduction in the bill reported by the House. The final bill fell within the limitations imposed by the House Budget Committee, which established spending ceilings but made no attempt to influence the substance of the appropriations bills affecting Forest Service programs. In the Senate, however, the routine functions of the Interior Appropriations Subcommittee and the Budget Committee ran headlong into one another. This began a series of false starts that got the Forest Service budget thoroughly caught up in several major budgetary struggles in which the Forest Service had no stake one way or the other. The Interior Appropriations Subcommittee's actions were consistent with the characterization of the Senate as an appeals court through which the agencies seek to have House cuts restored.

The Forest Service Budget as an Outcome of Bureaucratic Politics

A better understanding of the values and incentive structures of the many organizations that influence the Forest Service budget is valuable in explaining the outputs of each organization, but it reveals little about how these different outputs were reconciled to arrive at the final product, the annual budget for the Forest Service. The budget proposals that emerge from each participant are manifestations of differing priorities within each organization, alternative goals and policies that are to some degree in conflict with one another. In the environment of fragmented decision making and shared power that characterizes even the executive branch, these conflicts must be resolved through persuasion, accommodation and bargaining, "the pulling and hauling that is politics."[10]

Due to the nature of this environment, simply examining the budget proposals of each organization relative to an intermediate outcome like the President's Budget or the final outcome of congressional appropriations cannot reveal the dynamics of the bargaining and accommodation that have taken place, and care must be exercised in any interpretation of the interplay between any less than the full set of participants. An examination of the alternative budget proposals within the executive branch alone will understate the impacts of the bargaining that has taken place. The hierarchical structure of the executive branch permits the secretary of agriculture to state his or her preferences in the Department Allowance regardless of any unresolved conflicts with the Forest Service. OMB certainly states its own preferences in the President's Budget, unconstrained by any remaining differences in viewpoint with the secretary's office or with the Forest Service. But one of the critical factors underlying the sharing of power within the executive branch is the independent power base that each participant has in the Congress and in the court of public opinion. The lines that define the true power of any of the executive branch organizations do not follow the neat pyramidal configuration on the organizational charts; and for the Forest Service, the indirect lines linking it to key members and committees in the Congress may define its *de facto* power more accurately than the direct lines linking it to its executive branch superiors.

An additional precaution in interpreting the budget proposals of each organization as an accurate representation of its values and priorities is that, even at the organizational level, there is an element of strategic gaming in the budget proposals. To the extent that the general budgetary roles ascribed by Wildavsky to the various institutions of government are considered stable, each organization participating in the budget process couches its proposal to produce the desired outcome from the entire process, given every other participant acting on the budget request in accordance with expected patterns of behavior. To the extent that such strategizing and posturing is taking place, the funding pattern most desired by any one participant may not necessarily be the funding pattern that appears in its proposal. It can be argued, for example, that, knowing that the Congress generally acts to increase funding for the timber program over what has been requested by the Forest Service, the agency can place less emphasis on this program in its budget request than is actually desired. If the agency is working within a fixed overall budget ceiling imposed within the executive branch, such a strategy could pay off exceptionally well if the Congress opts to increase timber funding without correspondingly decreasing funding for the nontimber programs, in effect boosting the agency's overall budget. The danger in such a strategy is that the other participants may stray from their established roles and the agency may actually get what it asked for, finding itself without adequate funding for high priority programs. The departure of any participant from its expected role, especially in the Congress, can throw the agency and other participants into a panic.

Changes in the political and economic variables from one year to the next make it difficult to work out a budget strategy for individual programs or the details of bargaining in advance as an *organizational* task, that is, a task performed according to standard operating procedures. The time frame for intelligence gathering, proposal, counterproposal and decision is so short that the process soon becomes a function not of organizations but of small clusters of individual men and women bargaining as both leaders and representatives for their respective organizations. Among the organizations influencing the Forest Service budget, the key players each have a different relationship to their organization; their organizations act to both support and constrain them, in roughly equal proportions. The chief of the Forest Service, historically a career professional who has come up through the ranks of the agency, is as much a product of the Forest Service's highly developed system of indoctrination and behavior modification as any district ranger described by Kaufman in his classic study, *The Forest Ranger.*[11] The chief's highest priorities in the bargaining sessions are largely a function of Forest Service organizational values, and there are limits to what the chief can concede without risking the continued support and confidence of "Mahogany Row," the upper echelon of deputy chiefs, and of the regional foresters. The assistant secretary of agriculture, on the other hand, is constrained by the professional staff of the Office of the Secretary only to the extent that he or she fails to get adequate analysis of all of the possible alternatives. The support he or she receives is the mirror image of this--it is largely limited to the analysis of alternatives, with no particularly strong identification on the part of the professional staff with the individual goals of the assistant secretary. And at OMB, much of the *de facto* bargaining over the Forest Service budget is done by the professional staff (i.e., the OMB budget examiner for the Forest Service and the career deputy associate director) usually not involving the upper level political executives in any substantive way.

Among the congressional committees, most of the bargaining takes place directly among the appropriations subcommittee chairmen and ranking minority members, usually with the direct and personal involvement of the professional subcommittee staff members assigned to the Forest Service budget. Other negotiations may take place *sub rosa* between individuals from the Forest Service or the secretary's office with other members of the subcommittee or full committee regarding favorable amendments or report language to be offered during the mark-ups; but if this is done in an attempt to override a decision of the subcommittee leadership it is done at the peril of the agency or department. The relationship between the agency and the appropriations subcommittee leadership (and the subcommittee staffs) is a continuing one and is cultivated by the top Forest Service officials with exceeding delicacy and care. Any attempt by the agency to go around the subcommittee leadership would not be looked upon kindly, and the fallout would be only as far away as the following year's appropriations.

Explaining the outcomes of past Forest Service budgets in terms of the individual personalities involved in the bargaining and decision making introduces a vast array of variables. The individuals involved and the actions they take are, like the organizations they represent, a function of their history and the values they have evolved. One element of this history is the "baggage" the individual brings with him or her, the sensitivities to certain issues, commitments to various programs, and personal standing and debts with groups in the society.[12] Another element is the history of the *relationship* between the individual players. Political appointees may be quite transient, but other players in the process may deal with one another as individuals repeatedly over a period of years. It is especially within this element that the personality of the individuals becomes a key consideration, that "each player's basic operating style, the complementarity or contradiction among personalities and styles in the inner circles [become] irreducible pieces of the policy blend."[13] For example, when the RPA legislation was in Congress, the Forest Service Programs and Legislation staff, under then-Deputy Chief Max Peterson, worked hard for its passage despite opposition from OMB. The RPA, of course, prevailed and became law in spite of OMB's recommendation for a presidential veto. Shortly afterwards, an OMB career professional deeply involved in the evaluation of the RPA legislation, Donald Crabill, became the OMB deputy associate director responsible for the Forest Service budget; and Max Peterson became chief of the Forest Service. The difficulty the Forest Service has had getting RPA Programs approved by OMB, and the fact that Forest Service annual budgets have subsequently looked more like the OMB recommendations than the RPA Program budget targets, may be related in no small part to the continued enmity between these two individuals.

Although the individual occupying the position of assistant secretary has a relatively short tenure in this inner circle, his or her personality and predispositions can make a profound difference in how the other principals deal with one another. The assistant secretary in the Carter administration, Rupert Cutler, built a reasonably cooperative relationship with then-chief of the Forest Service John R. McGuire and with his successor, Peterson. Although predisposed to favor the non-commodity resource programs on the national forests, Cutler got along well with Congress, passively supporting the timber program and seeking additional funding for programs such as fish and wildlife, and recreation. On the other hand, Cutler's successor during the Reagan administration, John Crowell, quickly established a combative relationship with the Forest Service, then Congress, and in his final undoing, with OMB. Crowell had come to the appointment from a long career and high level position in the timber industry, and entered with an undisguised agenda to change everything that was, from the timber industry's perspective, wrong about the Forest Service.[14] Although such an activist approach was unusual in this particular position, it is not unusual among federal political appointees more generally.

> With less time to work for their goals and with other interests to pursue, in-and-outers hope for quick results. . . . Their view is shaped by the shared images held by various subgroups within society and by the images prevalent in the professional groups with which they have associated prior to entering the government and to which they may return. These outside reference groups continue to shape their perception of the national interest.[15]

His focus was primarily on raising the national forest timber harvest level to achieve the full biological potential yield, and on preventing the setting aside of any additional areas of productive national forest land as congressionally-designated wilderness areas. At one point in his tenure, Crowell threatened to fire the chief of the Forest Service for not presenting a budget alternative with a sufficiently high timber harvest level (which would have made Peterson the first chief to be fired since Pinchot). Crowell's adversarial relationship with the chairman of the House Interior Appropriations Subcommittee, Congressman Sidney R. Yates, is regarded as a major factor in the subcommittee's suddenly departing from its established role and aggressively seeking reductions in the Forest Service timber program budget. The Forest Service attempted to restabilize its relationship with Yates under Crowell's more moderate successor, Peter C. Myers. But when Myers was promoted to Deputy Secretary he was succeeded by George S. Dunlop, an assistant secretary so contentious and irritating to the appropriations committee that long-time chairman Jamie L. Whitten used an appropriations bill as the vehicle to entirely eliminate the position of assistant secretary for natural resources and environment--and Dunlop along with it--from the Department of Agriculture.[16] This rather stormy recent history has limited the success of the Forest Service in bargaining with the appropriations committee leaders and any real rapprochement will probably have to wait until the appointment of a new, and more conciliatory, assistant secretary (assuming the appropriations committee reestablishes the position). Clearly, the personality of the assistant secretary, and his or her deportment of that office toward Congress, can make the individual as much a political liability as an asset, not just for the Forest Service but for the administration itself.

The technical competence and forcefulness with which the assistant secretary assumes the position must also be a factor in evaluating the role of bargaining. Political positions of this level throughout the federal government are often filled by leaders from the private sector who accept the responsibilities not to gain power or to pursue a strong personal agenda, but out of a sense of *noblesse oblige*, taking time out of their careers to perform a civic duty.[17] In the position of assistant secretary for natural resources and environment overseeing the activities of the Forest Service, Rupert Cutler was the first individual to have a strong agenda for the agency. This was partly due to the narrowing of responsibilities for the position, as discussed above, and partly due to Cutler's own background as a former executive of a national conservation organization. The greatest departure from this general pattern, however, was John Crowell. Not only had Crowell's timber industry background given him a strong personal agenda, it gave him a better knowledge of the inner workings of the Forest Service, and its budget process, than any assistant secretary up to that time.

Though largely overlooked in Allison's models of bureaucratic behavior, the role of the highly motivated individual public policy entrepreneur becomes significant at the secretary and assistant secretary levels. Had it not been for the economic problems that plagued the timber industry throughout Crowell's tenure as assistant secretary, his combination of detailed knowledge and strong motivation may have resulted in significant amounts of public resources being deployed or redeployed toward ends that corresponded to his very personal vision of societal needs.[18] In concentrating on the way the influence by executive bureaus, Congress, or interest groups tend to transform well-stated policy into either routine outputs of institutions or the negotiated product of several institutions, bureaucratic politics theorists have in some cases underestimated the importance of interests, values and strategies in either promoting or impeding change.[19]

The strength of Assistant Secretary Crowell's personal commitment to increase the national forest timber harvest level can be seen in his decisions to attempt to do so against the advice of the Forest Service and the OBPA staff in his own office, and against contrary trends in important economic indicators. Crowell's understanding of the bureaucratic processes can be seen in his unprecedented move to cut off, or at least constrict, the informal lines of communication between the Forest Service and the Interior appropriations subcommittees. Although the Forest Service technically is prohibited from revealing the elements of their own budget request directly to the Congress, the agency usually finds a way of getting the information before the chairman through orchestrated congressional requests for information or some other such circumvention. The differences between the agency's request and those of the secretary or OMB are usually exploited by the appropriations subcommittee to the benefit of the Forest Service. Upon taking office, however, Crowell specified that no agency request was to be printed or submitted to him until the details had been worked out with the Forest Service beforehand, so that the agency request would itself already reflect the agreement reached between the Forest Service and the secretary's office. This is largely the reason for the extraordinary consistency between the agency request and the department allowance for fiscal years 1984-1986. The proposed timber harvest levels for fiscal years 1984 and 1985 were precisely the same, and what small difference there was in the FY 1986 request reflected the mounting congressional pressure on the Forest Service to recognize the current economic trends and provide them with a realistic proposal, notwithstanding any "interference" from the assistant secretary.[20]

Ironically, the intensity of Crowell's commitment, and the subsequent narrowing of perspective that it engendered, may have been much of the reason for his failure to achieve more of his personal agenda. In a review of his four years as assistant secretary, Crowell expressed his continuing surprise at "the lack of collegiality among the presidential appointees;"

> I had the perhaps naive notion that we were a team which was assembled by the president . . . to carry out some basic objectives upon which we all were philosophically agreed. . . . A pervasive influence affecting budgeting and policy initiatives within my purview was my constant awareness of the need to do something about the growing federal deficit and . . . I was in charge of one of the very few agencies of the federal government having potential for making money for the federal Treasury.[21]

OMB, however, did not share Crowell's unshakable certainty that the national forest timber program was a net revenue generator for the federal government and that more timber harvesting necessarily implied greater net revenue. OMB's reduction of the harvest level in the Department Allowance by 2 BBF in FY 1986 was not based on any particular aversion to timber cutting, but on OMB's strict adherence to marginal analysis and the philosophy of economy and efficiency. Crowell's long background and continuing close association with the timber industry continued to shape his perception of the national interest. Not sharing this particular "reference group," OMB simply had a different perception of what was in the national interest. Crowell apparently never recognized this:

> The Office of Management and Budget serves a real and needed purpose, but its handling of the proposed 1986 budget starkly demonstrates how no one super-agency should be running or trying to run the government or dictate policy to major departments and agencies. The chance to do something realistic about budget deficits with leadership from the Administration has largely been lost because of the dominant role OMB insists upon taking. . . . In too many cases, OMB is the tail which wags the dog.[22]

Frustration with OMB is not uncommon among appointed officials elsewhere in the federal government. In a 1985 survey of former presidential appointees by the National Association for Public Administration, more than one third asserted that dealing with the federal budget process was the most difficult aspect of their jobs.[23] In terms of bureaucratic politics, however, some measure of responsibility for Crowell's adversities must rest with his failure to respect the unspoken but palpable caste system that exists among presidential appointees. His inability to obtain the representation of Agriculture Deputy Secretary Lyng in his appeal to OMB, a function of the same structural factors that give the Forest Service its independence and autonomy within the Department of Agriculture, forced Crowell to step outside the protocol to command OMB Associate Director Khedouri's attention directly. His attempt to do so through a form of bureaucratic extortion almost guaranteed that his appeal would not receive favorable treatment by the political executives at OMB. His resignation followed a few short months afterwards.

Three conceptual models have been presented to arrive at a better understanding of the forces driving the decisions that emerge as the annual budget for the Forest Service. These different approaches are not meant to be used alternatively, but in concert with one another. Each approach takes a different perspective, and illuminating the subject from several different angles simultaneously decreases the likelihood that, like the blind men describing the elephant, a narrow sensing of one part will give rise to misleading conclusions regarding the shape of the whole. From the Rational Policy Model perspective, there is no doubt that the overarching need to control the budget deficit through reduced federal spending significantly influenced many of the individual decisions on the shape of the FY 1986 Forest Service budget. The threat of awakening the sequestration provisions of the Gramm-Rudman-Hollings Balanced Budget Act kept the need to reduce current budget outlays constantly in the forefront and roughly defined the boundaries of debate on the details of every agency's budget. Each organization within the executive and legislative branches of government with influence over the Forest Service budget further defined the boundaries of debate by reacting to the duress of budget constraints, each according to its own organizational goals, values and standard operating procedures. And in the final reckoning, the results of bargaining, persuasion and accommodation among the individual players representing their organizations' values--and their own perceptions--were the final determinants of the collection of decisions that emerge as the federal government's approach to managing the national forests. How much of each individual player's actions is determined by his or her organization's outputs, as opposed to individual perception and personal agenda, varies from one organization to another.

More than any other participant in the Forest Service budget process, the actions of the Office of the Secretary of Agriculture seem to have been defined more by the latter than by the former. Whereas the organizational outputs of most of the other participants create a relatively stable environment in which each participant's actions can be anticipated by the others, the actions of the secretary's office can be a wild card. This can perturb the established equilibrium between the other participants and substantially redefine their working relationships with one another, at least temporarily. An activist assistant secretary with a thorough understanding and regard for the values and incentives that drive the other participants in the process can make substantial progress over time in achieving a policy agenda for the Forest Service via the budget process. An assistant secretary who fails to understand or regard such forces and proceeds from an adversarial stance may be able to effect some degree of change at the outset of his or her term in office, but can quickly be isolated by a sort of bureaucratic immune system. Once this disturbance to the system has been, in effect, neutralized, the participants gradually return to their stable roles and the system gradually returns to the established equilibrium. This can serve as a powerful lesson to anyone seeking to stimulate--or impede--changes in agency policy through the vehicle of the budget process.

NOTES

1. USDA Office of Budget and Policy Analysis, "Decision Paper for Deputy Secretary's Mark-Up," Internal Memorandum, November 1984.

2. This assessment took in more than just the short-term financial gain or loss. To the extent that imputed benefits from non-commodity resources can be measured, and the long-term costs and benefits of capital investments (such as roads) estimated, these are integrated into the analysis as well. See: Sample, V. A., "Improving the Linkage Between the RPA Program and the RPA Assessment Findings: The View From the Office of Management and Budget," in: Binkley, C. S., Brewer, G. D., and Sample, V. A., (eds.), *Redirecting the RPA: Proceedings of the 1987 Airlie House Symposium on the Resources Planning Act,* Bulletin No. 95 (New Haven, CT: Yale School of Forestry & Environmental Studies, 1988).

3. Personal Communication, USDA Forest Service, Washington Office, Timber Management staff, July 1988.

4. Robinson, Glen O., *The Forest Service: A Study in Public Land Management*, p. 8-9.

5. 34 Stat. 1271.

6. 35 Stat. 259, 260, 267 (1908) as amended; 16 U.S.C. 500, 671; 31 U.S.C. 534.

7. 78 Stat. 1089, as amended; 16 U.S.C. 532-538. Computation of 25 percent payments extended to purchaser road credits began with FY 1976 under an amendment to the 1908 act. See: 35 Stat. 251, as amended; 16 U.S.C. 500, 553; 31 U.S.C. 534.

8. Lyons, James R. and Knowles, Donald, "The Forest and Rangeland Renewable Resources Planning Act: Congressional Staff Perspectives on Implementation of the Planning Process," in: Binkley, C. S., Brewer, G. D. and Sample, V. A., (eds.), *Redirecting the RPA: Proceedings of the 1987 Airlie House Symposium on the Resources Planning Act*.

9. Editorial, "Let the Timber Industry Hire Its Own Cheerleaders," Lewiston *Morning Tribune*, Lewiston, Idaho, May 15, 1988.

10. Allison, *Essence of Decision*, p. 144.

11. Kaufman, *The Forest Ranger*.

12. Allison, "Conceptual Models and the Cuban Missile Crisis," p. 714.

13. Ibid.

14. Petersen, Philip V., "An Interview with John B. Crowell, Jr.," *Journal of Forestry*, Vol. 79 (1981), pp. 379-382.

15. Halperin, Morton, *Bureaucratic Politics and Foreign Policy* (Washington, D.C.: Brookings Institution, 1974). p. 89.

16. A new position within USDA was created for Dunlop, one which explicitly denied him any executive authority over the activities of the Forest Service. See: Sinclair, Ward, "Ready, Aim, You're Fired; Chairman Whitten Bags a Bureaucrat," *Washington Post*, January 1, 1988, p. A17.

17. Heclo, *A Government of Strangers.*

18. See: Lewis, Eugene, *Public Entrepreneurship: Toward a Theory of Bureaucratic Political Power* (Bloomington, IN: Indiana University Press, 1984). In analyses of the organizational lives of Hyman Rickover, J. Edgar Hoover and others, Lewis explores the role of entrepreneurial actors in public bureaucracies; in an alternative view to that of Huntington, Allison and other bureaucratic politics theorists, Lewis sees many policy outcomes as direct results of individual policy entrepreneurs as well as organizational outputs or the product of bureaucratic bargaining.

19. Pastor, Robert, "Continuity and Change in U.S. Foreign Policy: Carter and Reagan on El Salvador," *Journal of Policy Analysis and Management*, Vol. 3, No. 2 (1984), pp. 175-176. See also: Hill, Jeffrey, "Why So Much Stability: The Role of Agency Determined Stability," *Public Choice*, Vol. 46, No. 3, pp. 275-287.

20. Personal Communication, USDA Forest Service, Washington Office, Timber Management staff.

21. Crowell, John B., "Would I Do It Again?" Speech before the Western Timber Association, Portland, Oregon, May 1985.

22. Ibid.

23. Brauer, Carl, "Tenure, Turnover, and Postgovernmental Employment Trends of Presidential Appointees," in: Mackenzie, G. Calvin (ed.), *The In-and-Outers: Presidential Appointees and Transient Government in Washington* (Baltimore: Johns Hopkins University Press, 1987), p. 185.

FOREST SERVICE BUDGET EXECUTION

7

Forest Service Budget Execution for Fiscal Year 1986

The conference report accompanying the final appropriations bill contains the Congress' detailed spending directives for the Forest Service for the fiscal year and can be exceedingly specific. For FY 1986, the conference report to accompany the year-long continuing resolution, House Joint Resolution 465 (House Report No. 99-450), contained a great deal more detail than just the budget line items that appear on the budget proposal for the Forest Service.[1] This is especially true for construction projects and land acquisition. For example, within the total construction appropriation there were specific amounts directed to a road in a particular recreation area in Alabama, a trail on Mt. Saint Helens, a visitor center in North Carolina and a sheep crossing bridge in Arizona. Specific amounts for land acquisition were directed generally to projects like the Appalachian Trail or programs like habitat consolidation for endangered species. Individual acquisitions were also specified, such as a right-of-way in Idaho that the Forest Service claimed it already owned. The Forest Service was, in effect, directly ordered by Congress to drop the defense of its ownership rights and pay $4 million to the Idaho constituent.

For some national forest resource programs, a portion of the total appropriation for that program is directed to a particular project, such as the $240,000 from the appropriation for the soil and water program that was to be used for a National Academy of Sciences study of Mono Lake in California. In other instances, the Congress will direct a specific expenditure on some particular project but not specify where or how the Forest Service is get the funding, so it must be taken from somewhere within the overall National Forest System appropriations. For example, the agency was directed to spend $900,000 to remove "slash and cull logs from the Bull Run, Oregon watershed to preserve water quality and reduce fire hazards" but it was up to the agency to extract these funds from the timber program, fire program, soil and water program or some combination of the three. Congress uses appropriations to specify its priorities when they differ from those of the agency; for example, the research station at Sewanee, Tennessee had been repeatedly slated by the Forest Service to be closed--but it remained open for as long as the senior senator from Tennessee was majority leader in the Senate. Congress uses appropriations to set administrative policy; an administrative initiative to make land management more efficient through an exchange of lands between the Forest Service and BLM was effectively crushed by Congress' prohibiting the expenditure of any federal funds by either agency to accomplish this.

Finally, Congress uses appropriations to set national forest resource output goals, but only to a limited extent. For FY 1986, the conference report specified a timber sale level of 11.4 billion board feet (BBF) nationwide. Congressional directives were extremely concise for Oregon and Washington, where a total of 4.9 BBF was to be sold, 2.45 BBF of which was to come from reoffered timber turned back to the Forest Service under the terms of the Federal Timber Contract Payment Modification Act, which allowed timber purchasers to get out of contracts they had overbid and purchase the timber again at a lower price. Within the overall national requirement, timber sale levels were also specified for Montana and Idaho. For FY 1986, as for most other years, Congress specified no resource output targets for any other resource program other than timber.

Allocation of Funding from the Forest Service Washington Office to the Regions

Perhaps in order to compensate for the chronic delays in the congressional appropriations process, Forest Service budget allocation begins internally soon after the release of the President's Budget in January, nearly nine months before the start of the new fiscal year. This stage of the process, resulting in a specific "initial allocation" to each region and national forest, begins with an appraisal by the Program Development and Budget staff in the Washington office of the major differences between the President's Budget and the original Forest Service budget request. The Forest Service budget request itself is an important written agreement between the chief and the regional foresters, representing a great deal of compromise and adjustment to the needs and priorities of the administration and of the chief, and it is heavily relied upon as a guide to allocating both the tentative and actual funding. Accompanying the funding allocation is a resource output target assignment for nearly every resource program. Some of these targets, especially those for the timber sale level, are the specific targets assigned by OMB. Other resource output targets, especially those for non-commodity resource programs such as recreation or soil and water protection, are proposed by the Forest Service itself in the detailed budget justification, the "explanatory notes," presented by the agency to the congressional appropriations committees. With the exception of the timber sales level, these resource output targets are seldom specified in any subsequent congressional appropriations document. Nonetheless, once they have been presented to Congress in the budget justification, with all but the timber sales level usually left unchanged, they are considered by the Forest Service to have the imprimatur of Congress. The targets are solemnly passed down through the agency as immutable imperatives at least as important as the appropriations themselves.

The tentative budgets for each region are usually completed and transmitted to the regional offices in late May or early June. This is the beginning of the field season for most national forests, and the regional budget staffs have several months in which to evaluate their tentative budgets and output targets for the following fiscal year in light of developments during the current field season. In early August, the regional foresters are given the opportunity to meet directly with the chief and the Washington office PD&B staff for further adjustments in the initial allocation. This usually takes the form of a meeting in which additional funding requests (or in some cases, requests for a lower level of program funding, together with a lower resource target) are presented in writing with a brief narrative to justify the request. Prior to the meeting with the chief, each request for reconsideration is submitted to the Washington office where it is reviewed by the director of the Washington office staff in that particular resource program (i.e., a request for additional reforestation funding would be reviewed by the Washington office timber staff). The staff directors' comments and recommendations are considered by the deputy chief for the National Forest System who then makes his or her written recommendation to the chief on each request. The chief generally gives one of four responses: approval, assign a high priority if additional funding becomes available, assign a low priority if additional funding becomes available or disapproval.

What the major issues were during FY 1986 in the regions where the case study national forests were located is suggested by these excerpts from the requests for reconsideration made by the regional foresters and the responses recommended by the Washington office staff to the chief.

Eastern Region

1. Request: Raise the timber target by 36.2 million board feet (MMBF) and increase funding by $371,000 due to increased timber demand in the Lake States. Staff recommendation: No additional funds--shift funding among other programs within region.

2. Request: Increase recreation funding $626,000 in order to meet assigned targets at minimum standards to avoid having to close sites and reduce personnel. Staff recommendation: No change.

If additional funding becomes available, give Region 9 an increase consistent with all other regions.

3. Request: Wilderness protection and administration will be severely reduced as a consequence of the (OMB) 31 percent cut in funding; request increase to provide minimal wilderness administration and prevent heavy negative impacts in Boundary Waters Canoe Area. Staff Recommendation: No change. If additional funding becomes available, give Region 9 an increase consistent with that of other regions.

4. Request: Initial allocation for fish and wildlife represents a significant reduction in targets and funding from historic program levels. This will be a setback to consistent and timely wildlife habitat improvement on timber sale areas; request additional $294,000. Staff recommendation: Increase target and funding as requested.

5. Request: Budget cuts will make it impossible to complete many outstanding land acquisition cases; request restoration of $308,000 to complete them. Staff recommendation: Region 9 has heaviest case load in the entire NFS; authorize additional funds.

Southern Region

1. Request: $300,000 to complete an air quality study required by Congress in the Virginia Wilderness Act; the need was not known when 1986 budget was developed. Staff recommendation: Concur with request.

2. Request: Recreation funding insufficient to meet assigned targets. Staff recommendation: No change; low priority for additional funding.

3. Request: Trail maintenance funding insufficient to meet assigned targets. Staff recommendation: No change; low priority for additional funding.

4. Request: Additional $647,000 for wilderness management; wilderness acres have increased 190 percent and funding has decreased 14 percent. Staff recommendation: No change; low priority for additional funding.

5. Request: $50,000 crucial for completion of fisheries habitat study affecting several forests. Staff recommendation: High priority for additional funding.

6. Request: Consolidation of five requests for a total $5.7 million reduction in both trust fund and appropriated funding for timber road construction, based on reduced need. Staff recommendation: Concur.

7. Request: $100,000 increase in funding for cultural resource inventory to meet legal requirements. Staff recommendation: Low priority for additional funding.

Northern Region

1. Request: $308,000 (48%) increase for Special Use administration; current funding is insufficient to meet public demands. Staff recommendation: Region should reprogram funds from other uses to cover shortfalls.

2. Request: Budget reductions will make the region unable to meet minimum requirements for the protection of wilderness resources or the health and safety of visitors; request funding at least at FY 1985 level. Staff recommendation: Low priority for additional funding.

3. Request: Cuts in recreation will increase the backlog for reconstruction and increase existing resource degradation request continuation of current year's funding level. Staff recommendation: Low priority for additional funding.

4. Request: Wildlife funding is inadequate to meet commitments under Grizzly Bear Recovery initiative; additional $175,000 needed. Staff recommendation: High priority for additional funding.

Southwestern Region

1. Request: Additional $496,000 needed for wilderness management to meet the requirements of Arizona Wilderness Act. Staff recommendation: Low priority for additional funding.

2. Request: Additional $630,000 needed for cultural resource management to comply with the terms of an out-of-court settlement of litigation. Staff recommendation: High priority for additional funding.

Pacific Northwest Region

1. Request: Increase of $2.1 million needed in recreation funding; current allocation will require closure of 120 developed sites. Staff recommendation: Low priority for additional funding.

2. Request: Need increase of $975,000 for wilderness management or a reduction in the assigned target such that 2 million acres are reassigned to less-than-standard maintenance level. Staff recommendation: Low priority for additional funding.

3. Request: Approval to shift $668,600 from timber program to wildlife program; current wildlife program not implementable at this level of funding; no funding at all has been provided for old-growth wildlife management. Staff recommendation: Approve reprogramming.

4. Request: Increase of $200,000 in trail maintenance needed or reduce target from 8,711 to 7,900 miles; current allocation will not permit trails to be maintained to standard; deferred maintenance is resulting in a loss of investment. Staff recommendation: Low priority for additional funding.

5. Request: Increase of $140,000 in recreation construction funding for facilities and roads and reduce assigned targets for developed recreation site capacity. Staff recommendation: Reconsider after final recreation construction appropriations received.

6. Request: $10 million increase in funding for timber road construction to support the out-year timber sale target of 4.6 BBF. Staff recommendation: Would involve redistribution of initial allocation; no change at this time (" . . . most likely did not reflect true needs since no other unit submitted a request in this area."[2]).

7. Request: Additional $525,000 (29%) for trail construction or reduction of 140 miles (50%) in assigned target. Staff recommendation: Low priority for additional funding.

Table 7.1 gives an overview of the requests for reconsideration on the initial allocation submitted by the regional offices for the five case study national forests. The greatest number of reconsideration requests were in regards to the timber program. However, only one of these involved a request for additional funds and/or higher assigned targets. Five of the timber requests were for reductions in road construction funding in the Southern Region and most of the remainder were requests for higher authorizations for Knutson-Vandenberg trust fund expenditures on reforestation and timber stand improvement. By contrast, nearly all of the requests for other programs called for significant additional funding, lower assigned program targets or both. Major additional funding was sought by every region for recreation use management on developed sites and for the management of wilderness areas. Significant additional funding was sought by all

Table 7.1
Summary of Requests Submitted by Regional Offices for Reconsideration of Initial Allocation for Fiscal Year 1986

PROGRAM AREA	REGION 9 (FOREST 1)	REGION 8 (FOREST 2)	REGION 1 (FOREST 3)	REGION 3 (FOREST 4)	REGION 6 (FOREST 5)	TOTAL
TIMBER	6	12		8	26	52
RECREATION	2	6	2	5	17	32
FISH & WILDLIFE	1	10	4	8	23	46
SOIL & WATER	1	5		4	10	20
RANGE			2	1	3	6
MINERALS		1			2	3
PROTECTION			1	11	2	14
LANDS	2	2	3	2	9	18
FACILITIES	1	2		3	6	12
TOTAL	13	38	12	42	98	203

but one of the regions for fish and wildlife habitat management. Perhaps because of the increasingly apparent downturn in the market demand for timber during 1985, OMB's 10 percent cut in timber funding apparently caused no great consternation among the regions. On the other hand, the 16 percent cut in recreation and even the 3 percent cut for fish and wildlife seemed to have raised deep concerns in the field that the job could not be accomplished with the funds provided. Regional concerns over insufficient funding for soil and water protection seemed to focus on the authorized level of expenditure from the Knutson-Vandenberg trust fund, which is used mostly to protect or mitigate damage to soil and water resources in timber sale areas. The other area of greatest concern seemed to be in the lands program, particularly funding for the administration of non-recreation special use permits.

Since the regional foresters' negotiations with the chief over the initial allocation usually involved cutbacks from the agency request, which itself represented reductions from the aggregate of the regional budget proposals for most programs, these negotiations are a critically important step in the allocation of budget resources down through the agency. There is a palpable sense of competition among the regions as each regional forester attempts to justify a relatively greater allocation for some particular program in his or her region. These negotiations are closely watched by the forest supervisors, and more than one regional forester has returned home to a critique of his or her performance in adequately representing the special needs of the national forests in his or her region.

The process is repeated in each region as the regional forester must decide how funding and assigned targets will be allocated among the dozen or more national forests in the region. The negotiations between the regional forester and the forest supervisors over the initial allocation are important in that they provide a reasonable basis for the forest supervisors to move ahead with developing a plan of work for the upcoming year. The new fiscal year, which begins on October 1, is usually just about to begin as this distribution of the initial allocation is taking place in the regions. But the line and staff officers throughout the agency are well aware that Congress tends to fund most programs at a higher level than in the OMB proposal, and that there may be additional flexibility once an appropriations bill is passed. Most of the regional foresters' requests to the chief for reconsideration of the initial allocation are held until the signing of the appropriations bill. By this time, the regional foresters have gone over the proposed program of work for the upcoming fiscal year in the context of the initial allocation and have identified potential problems. With this in mind, they return to the Washington office to negotiate the allocation of the actual appropriations. Washington office resource program staff have been in close communication with their counterparts on the regional level all this time and are also aware of where the greatest needs lie. Given that the agency is already three months into the fiscal year by the time the appropriations are made, time is of the essence and the process for allocating what additional funding has become available is not nearly so lengthy or formalized as the process for the initial allocation.

The regional allocations are taken to the field and the initial allocation of funds and targets to each national forest is adjusted to reflect any changes made by Congress and new situations that may have become evident in the field. This "final advice" is usually received by each forest supervisor in late March or early April. Should a forest supervisor have an immediate problem with this final allocation, it can be appealed back to the regional forester who, at his or her discretion, can accommodate the request by readjusting the allocations among all the national forests in the region. A primary consideration for the regional staff from this point onward is that expenditures not exceed the appropriation for each resource program and that the eventual output accomplishments of the various national forests sum to the output targets assigned to the region. Throughout the remainder of the fiscal year, there is a constant exchange, primarily between the supervisors' offices and the regional offices, to shift both output targets and funds among the various national forests according to their ability to help the region as a whole come as close as possible to the overall fund allocations and assigned targets.

For FY 1986, the final advice contained an additional factor that significantly changed the budget allocation process to the regions. The Balanced Budget and Emergency Deficit Control Act of 1985 ("Gramm-Rudman-Hollings") contains a provision whereby the overall appropriations for most operations of the federal government must be reduced across-the-board to stay within annual deficit limits specified in the law itself. This sequestration is done across all federal programs without regard to the substance or needs of any individual program. For FY 1986, it was determined that the total funding appropriated by the Congress had to be reduced by 4.3 percent. This amount was withheld for all national forest programs and the remainder distributed through the Forest Service according to normal procedures. Several months later, the Supreme Court determined that certain provisions of the Gramm-Rudman law were unconstitutional and the law was declared void. Nonetheless, the 4.3 percent that had been withheld for FY 1986 remained in the Treasury and was never released to the various departments and agencies. The final allocation figures referred to throughout this document are, therefore, net of the 4.3 percent reduction.

Variations in Regions' Approaches to Fund Allocation

It is in these processes of funding and target assignment to the forests, and the subsequent dynamic adjustment of those assignments, that the greatest variation was found among the five case study national forests and regions for FY 1986. The most distinctive variation was the approach taken by each regional office to maintain the region's *flexibility* to respond to changed situations late in the fiscal year and still come out close to regional funding and target assignments. The other distinctive feature was the variation in *strategic behavior* on the part of the supervisors' offices in response to the differing philosophies and approaches at the regional offices.

Southern Region

Budget execution in the Southern Region is distinctive in two major aspects: (1) the relative centralization of the process at the regional office, and (2) the budget execution role played by each resource program staff director at the regional office. Unlike in most of the other regions studied, the regional office holds back a large portion of the final allocation rather than distributing it immediately to its national forests. Reportedly this is done in keeping with a long-standing conservative tradition of guaranteeing that the regional office will be able to cover all its own costs by withholding an ample margin for all programs at the start of the fiscal year. As the fiscal year unfolds, and the regional office is more certain of what its own final costs will be, a portion of these funds is released and allowed to trickle down to programs at the national forest level. A more important result, and the more likely motivation for this behavior, is that by retaining a substantial share of the program funds, the regional office also maintains a high degree of flexibility for directing those funds to the national forests best able to help the regional office meet its assigned targets within its spending limitations. The apparent philosophy is that it is easier to hold back the funds in the first place than to take away funds that have already been allocated to one forest and transfer them to another halfway through the fiscal year.

This relative centralization of budget execution at the regional office is reinforced by the role played by the staff directors for each resource program in the regional office. By way of background, prior to 1983, budget execution throughout the Forest Service was largely handled not by the line officers, but by the staff directors for each resource program. For example, once an appropriations bill was passed, the spending authority for the timber program became the dominion of the director of the timber staff in the Washington office, who then decided how the funding would be distributed among the timber programs for all the regions. The timber staff directors in the regional offices then had similar control of the distribution of timber funds among the national forests under their purview. Any adjustments made during the course of the fiscal year, such as the transfer of timber funds from one national forest to another with greater needs, were negotiated between the regional staff director and his or her counterparts at the supervisors' offices for the affected national forests. It was in the interest of these staff directors to get maximum performance, and the funds for each program tended to be allocated to units where they could get the largest output accomplished per dollar spent. In effect, this promoted overall efficiency by channeling funds to those forests that could show the lowest unit costs of production.

An opposing tendency, however, was the willingness of the timber staff directors to permit a certain amount of "leakage" of funds intended for the timber program to be used to support nontimber programs. While this could ostensibly raise the apparent unit costs of production for timber, the amounts were usually too small in comparison to the overall timber budget to make any noticeable difference. Moreover, this had a number of benefits for the timber staff officer: (1) as more of the nontimber activities came to be funded through the timber program, this swelled the timber budget and, in accordance with the conventional wisdom regarding bureaucratic behavior, brought added prestige, and (2) to the extent that funding for nontimber programs came, in effect, through the timber program (the extent was considerable on many national forests) the budget, and thus the power, of many nontimber programs became subjugated to that of the timber program. During this period, the timber staff directors were widely referred to internally as "timber czars." A management consultant to the Forest Service recommended in 1982 that this system be changed so that all budget execution decisions had to flow directly through the line officer at each administrative level. It was felt that the line officer for a particular unit should have more control over non-timber program priorities than the timber staff director. This change was adopted by the Forest Service to apply to budget allocations beginning with FY 1983.

On Forest 2, officials felt that this process that had been ostensibly superseded had never really been changed; indeed, several senior officials were unaware that the policy change had ever been ordered. The only manifestation of this change that was noted was that there is now an attempt by all program staff directors in the SO to coordinate fund allocations to the ranger districts through the SO's administrative officer, essentially the forest supervisor's budget officer. The reliance of the line officers on the generally more experienced and longer tenured program staff directors is so great that, in many cases, the program staff directors still have a large measure of *de facto* control over fund allocations down the line in their particular program areas. The direct communications network up and down the organizational levels within a particular program area is still extremely active and effective at keeping the program staff director at one level constantly up to date on the needs and capabilities of that program area on all the units at the next lowest level. Actual fund allocations must simply have the signature of the line officer rather than being made directly by the program staff officer.

The combination of these factors has resulted in a predictable behavioral adaptation among the national forests in the Southern Region. In this region, there are "rich" forests and "poor" forests, a dichotomy based largely on the size of the timber program relative to the other resource programs. Under the Knutson-Vandenberg Act of 1930, what is in effect a portion of the receipts from the sale of national forest timber may be retained by the Forest Service in a trust fund to support timber sale related reforestation, timber stand improvement, fish and wildlife habitat management and soil and water protection.[3] These K-V funds can constitute a significant addition to the appropriated nontimber program funding on national forests with sizable timber sale programs. National forests without large timber sales programs also lack the additional funding from K-V deposits and thus have to get by on only what is directly appropriated. For national forests in the Southern Region, the budget allocation may be rather slim, but the forest supervisors and their program staffs know that there is a good chance that additional funding will become available later in the fiscal year, as the regional office relinquishes its hold on funds retained there. The poor forests are more willing to gamble on the availability of such late-year funding and set their programs at higher levels than can be supported with the final allocation alone. They then compete aggressively with the rich forests for whatever additional funding does become available in order to cover their planned activities.

As federal employees, Forest Service field officials bear significant personal risk in programing expenditures in excess of the funding they have in hand. The Anti-Deficiency Act specifies substantial fines and jail terms for federal employees who overspend their appropriations.[4] Forest 2 is a poor forest and officials there consider themselves much more willing to make the necessary effort to capture any late-year funding that trickles down than their colleagues on rich forests elsewhere in the region. Officials at the SO and the ranger districts put in many long hours in the last eight weeks of the fiscal year developing project plans to legitimately absorb any additional funds. For poor forests, some of the best funding opportunities of the entire fiscal year come in the final two months in a process described by Forest 2 officials as "a perennial scramble." Still, there is no guarantee that the books will balance, and as the fiscal year draws to a close, the communications between program staff directors at the different organizational levels become extremely active as officials try to determine exactly how much funding will be made available and how it will be allocated among the field units. The SO and even the ranger districts fine-tune their spending plans so as to not go over the limit, but also not to "leave anything on the table" (Forest 2 officials describe this process as "exhilarating--like jumping out of a plane and not being entirely certain your parachute is going to open"). The ranger districts work closely with the SO to maintain a series of "cutouts" that can be eliminated on short notice if prospects for additional funding turn suddenly bleak.

The use of a budget officer at the SO to coordinate allocations to the districts significantly improves the efficiency of this late-year scramble. When the program staff directors at the SO dealt directly with the ranger districts, they might call on short notice and ask, for instance, whether or not the district could usefully absorb an additional amount for wildlife habitat improvement. The recreation staff director may then call a week later inquiring after a place to put some additional recreation funding. By having the program staff directors coordinate all additional funding through the budget officer, the SO is able to "package" the funds so that they can be directed to integrated resource management projects that can most effectively and efficiently utilize the particular *mix* of program funds that has become available. This also gives the forest supervisor considerably more control over funding, even in the final mad rush, and allows it to be directed according to the supervisor's priorities.

Southwestern Region

FY 1986 fund allocation in the Southwestern Region was much more decentralized than in the Southern Region in that there was a concerted effort to get all the funding possible down to the national forest level as soon as possible in order to let forest supervisors know exactly what budget resources they had to work with. In fact, it was the regional office that was ostensibly overextended just after the final allocation, not the field offices. However, budget execution still became quite centralized by the restrictive manner in which the funds were allocated, by program, to each national forest in the region.

During the initial allocation phase, the regional budget staff developed a very complicated methodology for allocating funds among the forests, based on a detailed quantitative analysis of past funding and management activity in each resource program. Multiple regression and linear programming techniques were used to examine program costs among the various forests in the region and to predict funding needs in FY 1986 on the basis of a series of independent variables referred to as "complexity factors." For timber sale harvest administration, for example, the independent variables included: timber volume offered, proportion of offered volume in sawtimber and pulpwood, proportion of volume to be logged using high-lead cable yarding systems, average size of timber sale, number of timber sales, and number of fuelwood permits. For many of the nontimber resource management programs, funds were simply allocated on the basis of the previous year's allocation. In the case of construction activities, such as the construction of recreation facilities, funds were distributed according to "regional priorities as approved by Congress." The quantitative analysis technique was portrayed as an objective process producing "an equitable mix of functional [i.e., resource program-specific] dollars among the Forests." However, in situations where this program specified a fund allocation that deviated significantly from past funding levels for an individual forest, a "three percent rule" was used; in order to prevent any major increase or decrease for any individual forest (in the interest of equity), deviations were limited to no more than three percent above or below what the final appropriation had been for FY 1985. If this produced a solution in which the total for any program exceeded the regional allocation, the entire distribution among the national forests was reshuffled to bring the total within the specified limit. Although concepts similar to those used in linear programming were employed, no such model was actually developed to accomplish this analysis.

The reaction from the supervisors' offices to this rather mechanistic decision process was to assert generally that important factors had been either inaccurately quantified or left out entirely and that, had they been included, particular forests would have received a larger allocation. There was a great deal of cross-checking to see what the other national forests got, and claims that allocations on neighboring forests were excessive because their situation was not nearly as complex. Forest 4 responded that a number of additional complexity factors should have been considered in its allocation, especially its relatively greater workload in the recreation and lands programs. It was pointed out that a logical extension of the prescribed course of action over a period of several years would produce a result that was clearly irrational and detrimental to the field officials' ability to adequately protect and manage the resources. Forest 4 was also well along with its forest plan at this time and asserted that the mix of funding among programs was not consistent with the plan, that "the forest will not be able to develop the needed programs as identified in the land management plan because of the funding mix specified." These and other difficulties were the basis for a flurry of appeals from the Forest 4 supervisor to the regional forester. Most of these appeals were resolved through the traditional direct negotiations between the forest supervisors and the regional forester, and the analytical model for fund allocation was not used the following year.

Even though nearly all of the funding allocated to the region was passed down to the forests early in the fiscal year, the SOs still programmed activities in excess of appropriated funds, especially for such chronically underfunded activities as recreation, and soil and water protection. Some of the additional programming was based on requests for supplemental appropriations that had already been submitted to Congress. Some was in anticipation of an annual action by Congress, referred to generally as the "Pay Act," to permit increases in personnel costs primarily for salary increases. Unlike the regular Forest Service appropriations, Pay Act funds are not divided up by resource program. Thus, as they are distributed down through the organization, they constitute a pool of relatively unrestricted funding that can be used according to the priorities of the regional foresters or forest supervisors. In the Southwestern Region in FY 1986, these funds were largely funneled directly down to the forest supervisors to cover potential shortfalls. Beyond this, the balancing of expenditures with appropriations was relatively smooth and there were none of the last minute scrambles that characterized the end of the fiscal year on Forest 2.

Among the remaining three case study national forests, the budget allocation process in FY 1986 was much closer to that used on Forest 4 than that used on Forest 2. On both Forest 3 (Northern Region) and Forest 5 (Pacific Northwest Region), the budget proposal process had been highly automated and the mix of funds in the final allocations was not significantly different from what was proposed. The exception to this was the timber program, which was relatively lower for both forests, but caused little apparent difficulty because the demand for new timber sales was lower than had been expected when the budget proposals were being prepared. In instances in which the final allocation for a particular program was significantly below the absolute dollar level sought by the forests there was keen competition among the various forests for what funds had been appropriated to the region for that program. A resolution was negotiated between the forest supervisors and the regional forester, usually deferring largely to the regional foresters' priorities. For example, in dealing with a lower than expected allocation for the recreation program in the Northern Region, the regional forester's first priority was the health and safety of recreational users of the Bob Marshall Wilderness in western Montana and the fifteen national forests in the region had to compete for the remainder to cover other portions of the recreation program. In the spirit of decentralized management, most of the funding was quickly passed through to the national forests with only enough held back to cover the actual program and overhead costs anticipated for the regional offices. The exception to this for both the Northern and Pacific Northwest regions is construction funding, which was withheld at first and then funneled to particular projects according to proposals submitted by the forests and decisions made on the basis of regional priorities.

On the continuum between Forest 2 and Forest 4, the FY 1986 budget allocation process for Forest 1 (Eastern Region) was much like that in the western regions but, because it also qualifies as a relatively poor forest in that it does not have a large timber program, it also displayed elements of the process described for Forest 2. The Eastern Region office did not engage in the practice of holding back significant amounts of program funding with the intention of further distributing it later in the fiscal year. Consequently, there was not the heavy reliance on large amounts of program funding becoming available near the end of the fiscal year. Nonetheless, where there had been chronic shortfalls in funding, particularly for recreation construction, the forest supervisor and district rangers extended themselves beyond what was provided in the final allocation in hopes of having an opportunity to compete for additional funding later in the fiscal year. Program staff at the ranger districts made every effort to develop integrated project plans that could be kept "on the shelf" in order of priority so that there would be no delay in submitting a proposal to the forest supervisor in the event that additional funding came through. As on Forest 2, however, the forest supervisor always knew where there were cutouts which could be eliminated on short notice to avoid overspending at the end of the fiscal year.

The dichotomy between Forests 2 and 4 is one manifestation of a continuing difficulty in the administration of decentralized organizations: the conflict between specialists concerned with a particular *function* over a number of areas (often over several administrative levels) and generalists concerned with several different functions within a given *area* over which they have been given authority.

> In government the architecture of the whole rests on two pillars--major function and area. In any government the major functions and subfunctions can be grouped into a few great clusters for coordinated direction by a top generalist. Through this device the virtues of functional specialization are maintained, but the mutual adjustment of specialties in the light of a major goal of government becomes possible. . . . Those familiar with this device know that functional groupings are never perfect . . . that the generalist supervising a cluster of functions may be unable to establish firm control. The problem of area and administration is [one of] reconciling the parts and the whole, of introducing coherence in an age of specialization.[5]

Until fairly recently, as noted above, the timber czars and other functional specialists held a great deal of *de facto* decision authority, both at their own administrative level and over those below. The decision to shift budget allocation authority entirely to the line officers can be seen as a measure intended to shift the locus of power in the field offices of the Forest Service from the staff specialists more to the line generalists, i.e., from function to area. As can be seen in the differences in process that remain between Forests 2 and 4, some regions have been able to bring about this shift of power more quickly than others. As Forest Service land management planning moves from functional planning to integrated multi-resource planning under NFMA, this "mutual adjustment of specialties in the light of a major goal" during budget allocation will have important implications for successful plan implementation.

Comparison of Proposed, Appropriated and Reported Budgets and Output Targets for the Five Case Study National Forests

To begin to assess how much of an impact the modifications in the field budget proposals actually had on forest supervisors' and district rangers' abilities to accomplish planned resource management activities, comparisons were developed for each forest showing both the funding and the output targets proposed by the field offices, those allocated through the appropriations process and those actually reported back to Congress (see Tables 7.2 through 7.6). The objective of these comparisons is to determine the degree to which significant changes in budget allocations resulted in a significant change in program outputs and target accomplishments. The proposed funding levels and targets are those appearing in the original field budget proposals submitted by each national forest to the regional office. In the case of Forest 4, the funding proposals were those developed by the Southwestern Region office since no proposal was developed at the national forest level; no accompanying resource output proposals were developed at the regional office for the individual national forests in the region. The assigned funding and targets were those shown in the final advice to each national forest, which updated the initial allocation to indicate actual funding allocations and assigned targets in the appropriations bill. Data on the reported expenditures were obtained from information submitted at the end of the fiscal year for each national forest, the "statement of obligations" for the fourth quarter of FY 1986, to the Department of Agriculture National Finance Center. The information on reported outputs was taken from the "management attainment reports" for each national forest, the records of target accomplishments that form the basis for the "Report of the Forest Service" submitted each year by the Forest Service to Congress.

The most salient finding from these analyses is that, for most of the national forest resource programs, there is no clearly recognizable relationship between variations in funding and variation in outputs. The reported output levels were found to be either largely insensitive to relatively major changes in funding or to vary inversely with the changes in funding, illustrating the limits on interpreting the data without further exploring the context of the programs and any extenuating circumstances. On Forest 1, for example, reported expenditures on the timber program were 24 percent less than had been requested, yet outputs were 5 percent *higher* than had been projected in the field budget request (see Table 7.2). Similarly, 18 percent lower than requested spending for recreation produced 4 percent higher outputs. But for the fish and wildlife program, a 40 percent cutback in spending resulted in a 77 percent shortfall in outputs; for the soil and water program, a 27 percent cutback resulted in outputs 77 percent lower than projected. The patterns were different for every forest studied. On Forest 2, a 1.6 percent increase

Table 7.2
Forest 1: Comparison of Requested, Assigned and Reported Budget Expenditures and Resource Program Outputs, FY 1986, by Resource Program.

Budget Expenditures

DECISION VARIABLE*	A BUDGET REQUEST	B FINAL ALLOC	% CHG A>B	C REPORTED EXPENDITURES	% CHG A>C	% CHG B>C
	[1982 DOLLARS IN THOUSANDS]					
TIMBER	1,491.1	923.9	-38.0%	1,134.8	-23.9%	22.8%
RECREATION	801.3	558.6	-30.3%	659.9	-17.6%	18.1%
FISH & WILDLIFE	249.1	136.2	-45.3%	148.1	-40.5%	8.7%
SOIL & WATER	166.1	123.8	-25.5%	121.8	-26.7%	-1.6%
RANGE	72.6	64.3	-11.4%	62.9	-13.4%	-2.2%
MINERALS	84.0	51.4	-38.8%	53.3	-36.5%	3.7%
PROTECTION	46.7	40.1	-14.1%	42.2	-9.6%	5.2%
LANDS	247.2	324.0	31.1%	309.9	25.4%	-4.4%
FACILITIES	357.5	262.5	-26.6%	317.0	-11.3%	20.7%
GEN'L ADMIN	897.5	805.8	-10.2%	807.3	-10.1%	0.2%
TOTAL	4,413.1	3,290.6	-25.4%	3,657.2	-17.1%	11.1%

Program Outputs

	UNIT OF MEASURE	X PLANNED OUTPUT	Y TARGET OUTPUT	%CHG X>Y	Z REPORTED OUTPUT	% CHG X>Z	% CHG Y>Z
TIMBER	(MMBF)	12.0	12.2	1.7%	12.6	5.0%	3.3%
RECREATION (1)	(M PAOT)	317.9	331.7	4.3%	329.4	3.6%	-0.7%
FISH & WILDLIFE	(ACRES)	606.8	136.2	-77.6%	137.1	-77.4%	0.7%
SOIL & WATER	(M ACRES)	21.4	5.0	-76.6%	5.0	-76.6%	0.0%
RANGE	(M AUM)	463.7	459.0	-1.0%	471.0	1.6%	2.6%
MINERALS	(CASES)	64.2	85.0	32.4%	0.0	-100.0%	-100.0%

(1) Recreation outputs are measured in terms of use capacity rather than actual use. The term PAOT denotes the number of Persons At One Time that can be accommodated by recreation facilities.

Table 7.3
Forest 2: Comparison of Requested, Assigned and Reported Budget Expenditures and Resource Program Outputs, FY 1986, by Resource Program.

Budget Expenditures

DECISION VARIABLE	A BUDGET REQUEST	B FINAL ALLOC	% CHG A>B	C REPORTED EXPENDITURES	% CHG A>C	% CHG B>C
	[1982 DOLLARS IN THOUSANDS]					
TIMBER	3,325.5	3,503.7	5.4%	3,379.7	1.6%	-3.5%
RECREATION	611.0	765.4	25.3%	864.7	41.5%	13.0%
FISH & WILDLIFE	324.0	177.9	-45.1%	191.1	-41.0%	7.4%
SOIL & WATER	195.0	232.4	19.2%	223.7	14.7%	-3.7%
RANGE	77.5	81.6	5.3%	82.7	6.7%	1.3%
MINERALS	151.0	112.9	-25.2%	114.1	-24.4%	1.1%
PROTECTION	556.4	534.9	-3.9%	551.9	-0.8%	3.2%
LANDS	293.6	290.9	-0.9%	336.5	14.6%	15.7%
FACILITIES	478.5	517.1	8.1%	528.1	10.4%	2.1%
GEN'L ADMIN	1,204.6	1,169.2	-2.9%	1,238.6	2.8%	5.9%
TOTAL	7,217.1	7,386.0	2.3%	7,511.1	4.1%	1.7%

Program Outputs

	UNIT OF MEASURE	X PLANNED OUTPUT	Y TARGET OUTPUT	%CHG X>Y	Z REPORTED OUTPUT	% CHG X>Z	% CHG Y>Z
TIMBER	(MMBF)	21.3	27.0	26.8%	26.6	24.9%	-1.5%
RECREATION (1)	(M PAOT)	1,945.0	1,675.0	-13.9%	1,675.0	-13.9%	0.0%
FISH & WILDLIFE	(ACRES)	620.0	540.0	-12.9%	540.0	-12.9%	0.0%
SOIL & WATER	(ACRES)	21.3	54.0	153.5%	54.0	153.5%	0.0%
RANGE	(ACRES)	353.0	235.0	-33.4%	235.0	-33.4%	0.0%
MINERALS	(CASES)	52.0	19.0	-63.5%	227.0	336.5%	1094.7%

(1) Recreation outputs are measured in terms of use capacity rather than actual use. The term PAOT denotes the number of Persons At One Time that can be accommodated by recreation facilities.

Table 7.4
Forest 3: Comparison of Requested, Assigned and Reported Budget Expenditures and Resource Program Outputs, FY 1986, by Resource Program.

Budget Expenditures

DECISION VARIABLE	A BUDGET REQUEST	B FINAL ALLOC	% CHG A>B	C REPORTED EXPENDITURES	% CHG A>C	% CHG B>C
	1982 DOLLARS IN THOUSANDS]					
TIMBER	10,335.9	7,594.9	-26.5%	8,391.8	-18.8%	10.5%
RECREATION	1,631.0	855.3	-47.6%	1,160.4	-28.9%	35.7%
FISH & WILDLIFE	306.0	296.0	-3.3%	340.8	11.4%	15.1%
SOIL & WATER	305.0	249.0	-18.4%	254.9	-16.4%	2.4%
RANGE	55.0	36.8	-33.1%	36.1	-34.4%	-1.9%
MINERALS	217.0	170.0	-21.7%	171.6	-20.9%	0.9%
PROTECTION	1,013.0	818.9	-19.2%	809.5	-20.1%	-1.1%
LANDS	567.0	301.4	-46.8%	1,006.5	77.5%	234.0%
FACILITIES	1,152.1	819.4	-28.9%	919.2	-20.2%	12.2%
GEN'L ADMIN	1,773.0	1,744.5	-1.6%	1,808.2	2.0%	3.7%
TOTAL	17,355.0	12,886.2	-25.7%	14,899.1	-14.2%	15.6%

Program Outputs

	UNIT OF MEASURE	X PLANNED OUTPUT	Y TARGET OUTPUT	%CHG X>Y	Z REPORTED OUTPUT	% CHG X>Z	% CHG Y>Z
TIMBER	(MMBF)	100.0	100.0	0.0%	104.3	4.3%	4.3%
RECREATION (1)	(M PAOT)	491.0	279.0	-43.2%	693.4	41.2%	148.5%
FISH & WILDLIFE	(ACRES)	214.0	200.0	-6.5%	310.0	44.9%	55.0%
SOIL & WATER	(M ACRES)	40.5	10.0	-75.3%	10.0	-75.3%	0.0%
RANGE	(M AUM)	6.2	6.0	-3.2%	6.4	3.9%	7.3%
MINERALS	(CASES)	79.0	43.0	-45.6%	35.0	-55.7%	-18.6%

(1) Recreation outputs are measured in terms of use capacity rather than actual use. The term PAOT denotes the number of Persons At One Time that can be accommodated by recreation facilities.

Table 7.5
Forest 4: Comparison of Requested, Assigned and Reported Budget Expenditures and Resource Program Outputs, FY 1986, by Resource Program.

Budget Expenditures

DECISION VARIABLE	A BUDGET REQUEST	B FINAL ALLOC	% CHG A>B	C REPORTED EXPENDITURES	% CHG A>C	% CHG B>C
	[1982 DOLLARS IN THOUSANDS]					
TIMBER	5,321.4	3,256.9	-38.8%	3,420.6	-35.7%	5.0%
RECREATION	510.1	566.6	11.1%	545.0	6.8%	-3.8%
FISH & WILDLIFE	152.0	170.0	11.8%	184.0	21.1%	8.2%
SOIL & WATER	326.8	364.0	11.4%	377.0	15.4%	3.6%
RANGE	507.3	446.0	-12.1%	446.0	-12.1%	0.0%
MINERALS	67.5	52.0	-22.9%	51.0	-24.4%	-1.9%
PROTECTION	749.5	682.0	-9.0%	689.0	-8.1%	1.0%
LANDS	197.0	155.0	-21.3%	153.0	-22.3%	-1.3%
FACILITIES	996.2	1,395.7	40.1%	965.4	-3.1%	-30.8%
GEN'L ADMIN	1,097.3	1,040.0	-5.2%	973.0	-11.3%	-6.4%
TOTAL	9,925.1	8,128.2	-18.1%	7,804.0	-21.4%	-4.0%

Program Outputs

	UNIT OF MEASURE	X PLANNED OUTPUT	Y TARGET OUTPUT	%CHG X>Y	Z REPORTED OUTPUT	% CHG X>Z	% CHG Y>Z
TIMBER	(MMBF)	51.9	44.0	-15.2%	44.0	-15.2%	0.0%
RECREATION (1)	(M PAOT)	560.2	516.3	-7.8%	568.3	1.4%	10.1%
FISH & WILDLIFE	(ACRES)	500.0	500.0	0.0%	500.0	0.0%	0.0%
SOIL & WATER	(ACRES)	340.0	500.0	47.1%	400.0	17.6%	-20.0%
RANGE	(M AUM)	122.9	123.0	0.0%	123.0	0.0%	0.0%
MINERALS	(CASES)	50.0	33.0	-34.0%	63.0	26.0%	90.9%

(1) Recreation outputs are measured in terms of use capacity rather than actual use. The term PAOT denotes the number of Persons At One Time that can be accommodated by recreation facilities.

Table 7.6
Forest 5: Comparison of Requested, Assigned and Reported Budget Expenditures and Resource Program Outputs, FY 1986, by Resource Program.

Budget Expenditures

DECISION VARIABLE	A BUDGET REQUEST	B FINAL ALLOC	% CHG A>B	C REPORTED EXPENDITURES	% CHG A>C	% CHG B>C
		[1982 DOLLARS IN THOUSANDS]				
TIMBER	19,537.6	12,186.3	-37.6%	12,763.5	-34.7%	4.7%
RECREATION	938.1	1,021.1	8.8%	1,252.8	33.5%	22.7%
FISH & WILDLIFE	349.5	514.0	47.1%	329.8	-5.6%	-35.8%
SOIL & WATER	372.0	254.1	-31.7%	620.0	66.7%	144.0%
MINERALS	95.7	89.0	-7.0%	104.6	9.3%	17.5%
PROTECTION	922.4	924.0	0.2%	914.6	-0.8%	-1.0%
LANDS	609.0	404.0	-33.7%	418.7	-31.3%	3.6%
FACILITIES	1,599.0	1,454.6	-9.0%	1,593.9	-0.3%	9.6%
GEN'L ADMIN	2,190.0	2,042.8	-6.7%	2,016.1	-7.9%	-1.3%
TOTAL	26,613.3	18,889.9	-29.0%	20,014.0	-24.8%	6.0%

Program Outputs

	UNIT OF MEASURE	X PLANNED OUTPUT	Y TARGET OUTPUT	%CHG X>Y	Z REPORTED OUTPUT	% CHG X>Z	% CHG Y>Z
TIMBER	(MMBF)	266.0	254.0	-4.5%	261.7	-1.6%	3.0%
RECREATION (1)	(M PAOT)	256.0	213.0	-16.8%	407.0	59.0%	91.1%
FISH & WILDLIFE	(ACRES)	198.0	20.0	-89.9%	20.0	-89.9%	0.0%
SOIL & WATER	(ACRES)	35.0	5.0	-85.7%	5.0	-85.7%	0.0%
MINERALS	(CASES)	250.0	107.0	-57.2%	128.0	-48.8%	19.6%

(1) Recreation outputs are measured in terms of use capacity rather than actual use. The term PAOT denotes the number of Persons At One Time that can be accommodated by recreation facilities.

in timber spending led to 25 percent higher outputs, but a 41 percent increase in recreation spending resulted in outputs 14 percent below what was projected (see Table 7.3).

The outputs for the timber program showed remarkable stability on all the forests despite the substantial reductions between planned and actual funding that occurred in most cases. The timber program for FY 1986 had experienced a series of reductions, first from OMB, then Congress (especially in the construction of timber roads) and finally from Gramm-Rudman, so that at the national level the overall funding was nearly 20 percent below what had been proposed in the agency request. This reduction was not spread evenly among the 156 national forests, with the result that all the study forests except Forest 2 suffered at least a 25 percent cut from what field officials had originally proposed. Forests 4 and 5 were cut even further, by 35 percent or more. Forest 2 alone among the forests studied was able to spend more for its timber program than had been requested, and only by 1.6 percent. However, this 1.6 percent increase in funding allowed Forest 2 to achieve a nearly 25 percent increase in outputs over what it had proposed. What is even more surprising is that all of the forests, despite the decreases in funding, were able to nearly meet, and in many cases actually exceed, the timber sale level originally proposed.

Not one of the national forests studied fell more than 1.5 percent short of the timber targets assigned by Congress and allocated through the regional offices. In all but one case, the forests accomplished slightly higher than the assigned timber outputs, despite a difference of 38 percent or more between proposed and allocated funding for at least three of the five forests. The timber targets assigned to each forest were very close to those the forest had originally proposed, with the exception of Forest 2. Forest 2 was assigned a timber target more than 25 percent higher than what it had proposed, and was the one forest to accomplish less (but only 1.5 percent less) than it was assigned. Forest Service officials point out by way of explanation that the agency was responding to pressure from the Congress. By the time the House and Senate appropriations committees convened to write the conference report for FY 1986 appropriations, the Congress was fully aware of the extent to which the timber program budget had been cut back; yet they explicitly assigned a specific timber output target for the National Forest System as a whole and to the Pacific Northwest Region in particular. Forest Service officials felt this left no choice but to "pull out all the stops" in order to achieve the assigned targets. A considerable volume of timber turned back to the government under the Federal Timber Contract Payment Modification Act in the Pacific Northwest and Northern regions required minimal additional preparatory work before being reoffered. Even with this advantage, however, timber sales already prepared but intended for sale in later fiscal years had to be offered during FY 1986 to meet the targets (requiring a correspondingly greater effort to meet timber sale targets for several subsequent years).

Recreation outputs, as measured by recreation use capacity at each forest, generally fared better than might be expected from the 5 percent drop between proposal and appropriations for the National Forest System as a whole. Recreation funds were also distributed unevenly among the studied forests, from Forest 3, funded approximately 29 percent below its request, to Forest 2, funded at more than 40 percent above its request (see Table 3.4). It seems counterintuitive that Forest 3 should show accomplishments 41 percent *higher* than proposed and Forest 2 should show accomplishments nearly 14 percent *below* what it proposed, the reverse of what happened with funding. This is not explained by variation within the elements of the recreation program; funding for recreation use management, trail maintenance, trail construction and recreation facility construction were all lower than proposed for Forest 3 and higher than proposed for Forest 2. One possibility is that the methods for estimating and reporting recreation use capacity differ significantly between the two forests, a point that will be discussed further below. Accomplishments relative to the proposals were between these two extremes for the other forests studied. Forest 1 spent 15.7 percent less than it proposed, but was able to exceed its proposed outputs by about 4 percent. Similarly, Forest 5 spent 33 percent more than it asked for but was able to accomplish a 59 percent increase in outputs (see Table 3.6).

One factor in the recreation program that Forest Service officials at every location felt was highly significant is the contribution from volunteers. The Forest Service operates two major volunteer programs associated with recreation. Senior citizen volunteers are often provided a temporary residence at developed recreation sites to greet visitors, collect fees and coordinate maintenance activities. Expenses for this program, which were over $200,000 in FY 1986 for Forest 2 alone, are appropriated through the Department of Labor and then transferred to the Forest Service. A larger volunteer program employing individuals of all ages provides a daily subsistence allowance from regular Forest Service program funds and is used for trail construction and maintenance and a variety of other recreation-related activities. The contribution of volunteers has become almost vital to the recreation program; on Forest 3, which made particularly effective use of such "human resource programs," nearly 40 percent of the recreation outputs reported for FY 1986 were accomplished by volunteers.

Performance in the fish and wildlife program showed similar variations among the study forests. Most of the forests, again, fared much better than might have been expected given their budget allocation. Forest 2 spent 41 percent less than requested but fell only 13 percent short of its own output goal. Forest 3 received 11 percent more than it proposed but was able to accomplish 45 percent more in outputs. Forest 1, on the other hand, was cut back 40 percent from its budget proposal but fell short fully 77 percent from its outputs. In a more extreme case, Forest 5 received only a 6 percent cutback, but accomplished only 10 percent of its proposed goal. Officials on Forest 5 attribute this to the diversion of most of the wildlife management personnel to studies and projects involving threatened and endangered species, especially a massive study of the habitat requirements of the Northern Spotted Owl done in conjunction with the forest planning effort in the Pacific Northwest Region. Such activities do not contribute to regular fish and wildlife habitat management targets.

For the non-commodity resource programs, as with the timber program, the more meaningful comparisons may be not those between proposals and expenditures, but those between allocations and expenditures. In many instances, the reported expenditure for a particular resource program varied, sometimes considerably, from the final allocation. Rather than implying that the national forests overspent their appropriations for these accounts, this often indicates that additional funds were allocated to that forest during the course of the fiscal year. This may reflect additional or supplemental congressional appropriations. In most cases, however, the differences are a result of the Forest Service's internal shifting of spending authority from one unit to another that continues almost to the end of the fiscal year. The law requires only that agency obligations not exceed its appropriations at the national level, and the difference between allocations and appropriations can differ widely among individual national forests. Data for FY 1986 shows that there is much less of a difference at the regional level, and even less for the agency overall.

The changes between dollar allocations and actual expenditures at the forest level, whether upward or downward, seemed to have little influence on the change between assigned targets and reported outputs. There were only a few instances in which reported outputs substantially exceeded those assigned, and even fewer in which the assigned targets were not at least met. In the vast majority of cases, there was less than 1 percent deviation from the assigned targets--and in many instances the variation was zero.

Other Factors in Interpreting the Comparisons of Planned Versus Actual Funding and Outputs

In order to gain a more comprehensive understanding of what is implied by this data, a series of interviews was conducted with Forest Service officials associated with each national forest studied. The officials interviewed included budget, administrative and resource program staff officers at the regional offices, supervisors' offices and ranger districts as well as present and past line officers for the national forests and ranger districts. Additional documentation related to the events of FY 1986 was available in most cases, but a great deal more was gained simply from the recollections of the officers and staff directly involved in the on-the-ground management of the forests themselves during and leading up to FY 1986.

Two major qualifiers were advised in drawing conclusions from the above comparisons, one regarding the "softness" of the methods for determining and reporting the various output measures, and another regarding differences between the mix of expenditures among resource programs indicated in the accounting records and that which actually occurred "on the ground." Many of the field officials indicated that, among the resource management programs for which there are specific appropriations, there are those with "hard targets" and those with "soft targets." The hard targets tend to be the tangible, directly measurable outputs of commodity resource programs: board feet of timber, number of cows or sheep grazed, mineral leasing permits issued. The non-commodity resources tend to have less quantifiable or less tangible outputs: the user capacity of a recreation site or the number of users over a dispersed area in the course of a year, or the number of acres for that soil and water quality that has been improved. The estimation of the capacity of a particular campground may seem fairly straightforward, once certain assumptions are made about the number of campers or picnickers in an average party using a developed automobile campsite over a season lasting a particular number of days. The capacity of a million-acre national forest to accommodate a variety of dispersed recreational uses becomes more problematic. Even the use that actually takes place is seldom directly monitored for most of the national forest recreational activities, with the exception of a small number of heavily used wilderness areas whose use levels are determined through visitor logs or by the issuance of permits. Annual recreation use figures are notorious among field officials for being based on "a horseback estimate" of increase or decrease from the previous year's level, a figure which itself was based more on a manager's rough sense of use than on any direct quantitative measurement. Output levels for soil and water improvements tend to be even less definable. The installation of a series of culverts along the main collector road in a mountain drainage may be considered to have improved soil and water protection for the entire area of the drainage, which could as likely be 2,500 acres as 500 acres. Similarly, a debris dam built in a stream may be considered to have improved many miles of fish habitat, and the improvement of forage in a small winter range area for elk may be considered to have improved overall elk habitat for thousands of acres of accompanying, but less critical, summer range on the surrounding mountainsides.

Whether or not a forest supervisor has met his or her assigned target for timber sale volume is quickly and unequivocally determined by direct, physical measurement; whether recreation or wildlife targets have been met is a matter of broad professional judgment, and is within the discretion and authority of the forest supervisor and the SO staff to determine. Given that the meeting of assigned targets is one of the most critical components of annual performance evaluations for the line officers at both the national forest and ranger districts, such broad discretion for determining whether or not non-commodity resource program targets have been met could be a problem. However, Forest Service field officials indicated that the primary focus of performance evaluations continues to be the attainment of the "hard targets" for timber and, to some extent, range and minerals. On most of the national forests, the district rangers in particular indicated that their overwhelming management concern relating to their own performance evaluations was to "get out the cut," that is, meet the annual timber sale volume targets assigned them.

Many also pointed out, however, a growing emphasis on the district rangers' abilities to deal with issues related to local use of the forest's non-commodity resources, especially those arising from public involvement in forest planning. Rather than focusing on the attainment of numerical targets for non-commodity resources, superiors tend to judge a manager's performance on his or her ability to "contain" issues at the lowest possible organizational level. Those managers who can effectively deal with issues at the local level and prevent them from "bubbling up" to where they must be dealt with by the forest supervisor are rewarded, especially if they can also meet their timber sale targets. Thus, while management activities related to the non-commodity resource programs are certainly important to the local manager, it is unclear that the numerical assigned output targets, and their precise achievement as reported on the management attainment reports, are meaningful as indicators of what was accomplished with a given budget allocation.

The validity of the comparisons between allocated reported program expenditures and outputs is further compromised by the revelation that a significant portion of the expenditures at the ranger district level are reported not necessarily as they were spent, but as they were *planned* to have been spent. Thus, while a significant portion of the funding allocated for a particular program may actually have been spent to achieve high priority objectives in another program, the accounting records would continue to indicate that the funds had been spent just as allocated. The full extent of this accounting on an "as planned instead of as worked" basis is only now becoming known even to the Forest Service itself and to Congress, raising an issue of public accountability that will be discussed below.

An examination of the budget execution and accounting procedures on the studied national forests, and especially on their ranger districts, suggests that these deviations between actual and reported expenditures are not so much a function of malfeasance on the part of Forest Service field managers as of conventions in the budget execution and accounting procedures that have evolved in the agency over a period of years. There are two aspects to the problem. The first is the intentional bending of accounting requirements by local managers in order to compensate for mismatches between the mix of program funds allocated and the mix needed to carry out planned integrated resource management activities. The second, and apparently more significant, relates to the complexities of translating as many as 26 different appropriations into hundreds of individual integrated resource management projects, each one subsuming funds from many different appropriations--and resulting in a considerable deviation between planned and actual spending of which the local manager may not even be aware.

Mismatches between the allocated mix of program funding and the mix needed to carry out the necessary integrated management activities occur not only because of the vagaries of the budget development process and its many participants, but also because of the natural limitations in the field managers' abilities to precisely predict their resource management needs as they are devising a budget proposal for two years in the future. Field proposals for the FY 1986 budget had already been submitted to the regional offices prior to the start of the FY 1984 field season. Managers' foresight is further limited by a few very large random variables such as forest fires and insect infestations that may occur in the interim and entirely redirect a ranger district's program of work for the year. Congress has made some allowance for this by granting the Forest Service authority to "reprogram" funds from the appropriation for one program to be spent on another, but the authority is quite limited. For the entire National Forest System, the chief has authority to reprogram up to $250,000 or ten percent, whichever is less, per program area.[6] The regional foresters have no such authority whatsoever, so that all reprogramming requests must be dealt with at the top. Any reprogramming needs that exceed this limitation must go to both the House and Senate appropriations committees for *prior* approval, that is, Congress must approve the reprogramming request before the district ranger may legally obligate funds based on the request.

Field managers note two major obstacles to their making effective use of the agency's reprogramming authority when funding mismatches occur. The first is timing. If local reprogramming requests cannot be dealt with by shuffling funds between forests or regions so that the agency as a whole stays within the $250,000 limitation, the request must be submitted to Congress, which means it must first be approved by both the secretary of agriculture and OMB. A national forest or ranger district typically does not receive the final advice on actual appropriations until the start of the field season, and thus their managers do not recognize a need for reprogramming until June or July at the earliest. The local managers' authority to incur new obligations is cut off sometime in August to prevent end-of-fiscal-year buying binges to exhaust their remaining spending authority in the weeks leading up to September 30. So unless a reprogramming request can be dealt with internally, and at a fairly low level in the organization at that, chances are slim that the approval will be received before the year-end spending authority cutoff. Those managers who reported having used reprogramming to adjust their budgets also indicated that they have had to take the personal risk that the request would eventually be approved and obligate the funds in advance of actual approval. Should this be done and the request be subsequently denied, the line officer could be subject to severe sanctions from within the organization and possibly from without as well.

The second important obstacle to field managers' use of reprogramming is the persistent perception that admitting to needing less money in the area that the funding would be reprogrammed from will result in lower funding in that area in the following year. None of the officials interviewed had ever actually had this happen to them, but this seems to be the conventional wisdom throughout the organization (and indeed throughout many organizations, public and private) and does not appear likely to be relinquished any time soon. However irrational, this aversion to asking for a funding reduction in a particular program area is a powerful one, and acts to reinforce the timing obstacle. Officials on one national forest reported being directed by Washington office officials to submit a reprogramming request following the discovery of accounting irregularities during a general management review by the Washington office. Only after delays and several repeat requests from the SO did the district rangers submit the necessary information to formulate the request, and once the request had been submitted to the regional office it languished still further. When it was eventually submitted to the Washington office it was handled within the chief's authority and approved. Nonetheless, this national forest has not submitted another reprogramming request since.

Given the almost institutionalized reluctance to use the formal means for correcting funding imbalances, field managers are left to their own devices to find informal means for doing so. Again, knowing that their superiors place a premium on their ability to contain issues at the lowest possible organizational level, line officers at the national forests and ranger districts often align budget allocations to management needs on the basis of rather generous interpretations of budget execution regulations and laws. For the resource management programs within the so-called "P&M" portion of the National Forest System budget (for Protection and Maintenance; this includes virtually all national forest management activities other than the construction of roads and facilities) almost all of the funding represents salaries and benefits for resource management personnel. Allowances formerly built into each program's budget to cover overhead were removed by congressional direction in 1982 so that all expenses for administrative offices, clerical labor and even the salaries of the line officers are handled separately in the general administration budget line item. The reported expenditures for each resource program therefore depend heavily upon how each salaried individual's time is allocated among the different program accounts. For all but the largest ranger districts in the Pacific Northwest and California, where ranger district personnel are highly specialized, staff officers typically have responsibilities for several different resource programs. If one of those resource programs is better funded relative to management needs than another program, it is no great problem to simply charge more time to the better funded program, regardless of how the time was actually spent. One item mentioned without fail on every national forest and ranger district visited is the chronic inadequacy of funding for the administration of non-recreation special use permits. In dealing with this problem, the doctrine of issue containment quickly comes into play. In the words of one staff officer with responsibility for the range, fish and wildlife, and lands programs on his district,

> when someone comes through the door and wants a permit, I can't just tell him the agency isn't adequately funded for me to spend two hours with him and to come back again next year. I just spend the two hours with him, which means I have two hours less to spend elsewhere, and I don't bother charging my time to an account I know doesn't have any money in it. The emphasis is on getting the job done, whatever it takes.[7]

The high degree of budget line item differentiation seems to break down in actual practice, a situation not uncommon among field organizations. The district ranger, as a generalist concerned with the entire array of functions within his or her geographic area of responsibility, must take the most flexible approach permissible under law, regulation, and the explicit and implicit policies of the organization.

> In a sense it would appear that functions must draw together and abandon some of the niceties of specialization when brought face to face with the ordinary citizen in his local setting.[8]

A different method of resolving the same problem involves the use of accounts with no fixed appropriation as a sort of safety valve. The costs for controlling forest fires are, of course, a function of how many fires actually occur and are therefore impossible to know in advance. Until recently, forest fire fighting costs were covered in a special supplemental appropriation from Congress made toward the end of the fire season. With the 1974 decision to begin the federal fiscal year on October 1, this was changed, so that now the agency simply makes whatever expenditures and obligations are necessary to control the fires, and the total amount is appropriated almost as a reimbursement in the next regular appropriations bill. This provides a sort of revolving fund that works well in all but the worst fire years, when a supplemental appropriation may still be needed simply to cover current obligations. On the poor forests, as defined above, line officers who program higher personnel expenditures than they are allocated find themselves in great difficulty near the end of the fiscal year when no additional funding has materialized. In order to avoid simply furloughing salaried professionals, a substantial portion of the district staff may be made available for fire fighting duties on other national forests, next door or across the country. For whatever time these individuals are on fire duty, whether or not they are fighting fires, they are paid from the fire fighting account, sometimes through to the end of the fiscal year. On national forests that typically have a significant number of forest fires, the line officer will intentionally "deficit finance" his other program personnel so that, when they are placed on fire duty, there are not substantial resource management funds intended for their salaries that will be "left on the table" and have to be turned back at the end of the fiscal year.

Another technique is to use fire fighting crews to accomplish work for other resource programs while still being paid from the fire fighting account. Because the crews must be close enough to their equipment and transportation to be readily deployed, however, their availability for field work is limited. A variation on this theme is the redefining of a task so that it can be legitimately accomplished by fire crews. One district ranger recalled being denied additional timber stand improvement funds to clear out understory trees killed in an unexpectedly heavy insect infestation. It occurred to him that these dead trees were probably a fire hazard as well, especially given the more than adequate funding that existed for fire-related brush disposal. A fire crew was assigned to the ranger for "fuels treatment" and, incidental to the elimination of a fire hazard, the insect infestation was controlled.

By far, however, the greatest majority of incidences of accounting "as planned instead of as worked" occur much less intentionally and often without agency officials even recognizing what has happened. Project work plans at the ranger district level are usually formulated as integrated resource management projects, that is, a single project may have a timber component, a wildlife component, a soil and water protection component and even a recreation component. These components relate to one another in different proportions for each individual project. The project work plans are prepared with an awareness of the past funding limits for each resource program and generally reflect ample funding for programs such as timber, and lower levels of funding for many of the non-commodity resource programs. When the national forests receive their final advice near the start of the field season, there is an effort to review the season's work plans to be sure they are not grossly out of line with the funding allocation received. Each integrated resource management project is assigned a numerical management code and, for the remainder of the field season, an array of different personnel who spend time on a particular project will charge some portion of their time against its management code. Often the individual district staff officials never see the original work plan, with its assumed distribution of funding by program. They see only the management code to which they are charging.

Early in the field season, there is a concerted attempt by district officials to coordinate with the district administrative officer, who oversees program accounting, so that any obvious shortfalls or overruns in a particular resource program can be reported to the SO. The budget officer at the SO can then coordinate with his or her counterpart at the regional office to shift those program funds around to other national forests and ranger districts in the region and keep every account reasonably in balance. Toward the end of the field season, however, this complex and time consuming process becomes overwhelming. Field crews rushing to complete projects before the onset of winter no longer coordinate as uniformly with their own district administrative officers, making it difficult for budget officials at the higher levels to connect units that will experience shortfalls for a particular program with units that will experience overruns in that same program. No longer able to keep up with exactly how much is being charged by an array of personnel to particular resource programs via project management codes, the district administrative officer finds late in the fiscal year that too much has been charged to some resource programs and not enough to others. Rationales are quickly developed for legitimately shifting recorded expenditures from oversubscribed programs to those with remaining balances. The next evaluation of the district ranger's performance rests heavily on the skill of his or her budget officer at this point. In the end, all accounts are made to balance. For FY 1986, the entire P&M portion of the budget (the total of all the commodity and non-commodity resource programs) for Forest 3, one of the largest and most challenging in all the region, balanced to *within 0.2 percent* of what had been allocated. These program-by-program expenditure figures, once they have been so balanced, are submitted to the supervisor's office and combined with the corresponding figures for the other districts. A similar aggregation takes place at the regional office and at the Washington office, with the validity of the figures questioned progressively less with each aggregation. Finally, the figures become the basis for the annual report of the Forest Service, submitted to a vigilant Congress soon after the end of the fiscal year.

The responses of Forest Service officials at every national forest and ranger district visited strongly suggest that the greatest single source of variance between reported and actual expenditures is the inconsistency between the program funding mix assumed in individual project plans and the way the field personnel actually spend their time on the ground. A single program staff officer is often involved with dozens of different projects, each with a different proportional mix of funding. Seldom is there any ready awareness on the part of field personnel of the proportional mix of funding assumed in a particular project plan and how consistent it is with the program mix of the hours that they--and many others--are charging to it. They are simply charging to a numerical management code. Moreover, there is often no means by which the district ranger or the administrative officer can double check whether the time being charged to a project by all the various program personnel, especially multidisciplinary staff, matches at all with that assumed in the program plan. Whatever errors exist in the original charges to individual management codes, however unwitting, are carried through the entire accounting system. This is the primary data base on which the entire accounting and reporting structure rests. Not only is there no means of correcting errors in which resource program a charge was directed to, there is no means of even knowing they are there.

All of this suggests that, with the exception of the timber sale volume, the resource output figures flowing up through the Forest Service Program Accounting and Management Attainment Reporting System (PAMARS) are not very accurate representations of what is actually being accomplished in the national forests, nor are the accounting figures very reliable representations of how funds are actually being spent. Perhaps the greatest control exercised by the Congress over Forest Service activities is in its assignment of the national forest timber sale level in each year's appropriations bill. And even in this it is impossible for the Forest Service to accurately inform the Congress what it costs to produce these outputs.[9] There would appear to be enough play in the system for the Forest Service to accomplish its highest-priority work in most resource program areas, with the exception of construction projects, despite minor mismatches in the mix of proposed and actual funding. However, the Forest Service as an organization places more emphasis on the accomplishment of its congressionally-assigned output targets than on the precise accuracy of its expenditure of and accounting for congressional appropriations by budget line item. All other things being equal, the outputs of resource programs with "hard" targets are more difficult to fudge than those with "soft" targets. As the pressure grows to provide more from the national forests with fewer dollars, it seems inevitable that the soft target resource programs will shoulder a disproportionate share of the decline in quality as well as quantity (how ever this is defined).

NOTES

1. U.S. Congress, House of Representatives, Committee on Appropriations, *House Joint Resolution 465, Further Continuing Appropriations for Fiscal Year 1986,* House Report No. 99-450 (Washington, D.C.: GPO, 1986).

2. USDA Forest Service, Washington Office, *Chief/RF&D Negotiations: FY 1986 Initial Allocation*, Unpublished internal working document, August 2, 1985.

3. Knutson-Vandenberg Act of 1930, 46 Stat. 527, as amended; 16 U.S.C. 576-576b.

4. Federal Anti-Deficiency Act, 31 U.S.C. 1351 and 1517. See also: Executive Office of the President, Office of Management and Budget, *Instructions on Budget Execution*, Circular A-34, Part III (revised August 26, 1985).

5. Fesler, James W., *Area and Administration* (University, AL: University of Alabama Press, 1949), pp. 121-122.

6. USDA Forest Service, *Program Development and Budgeting Handbook,* Forest Service Handbook 1909.13, at section 36.1. These instructions are based on congressional direction usually appearing in the most recent conference report on Forest Service appropriations.

7. Personal communication, USDA Forest Service, 1987.

8. Fesler, *Area and Administration*, p. 124.

9. Testimony of R. Max Peterson, in: U.S. Congress, House of Representatives, Committee on Appropriations, *Hearings on Appropriations for the Department of Interior and Related Agencies for Fiscal Year 1985* (Washington, D.C.: Government Printing Office, 1985).

8

Accountability and the Forest Service Budget Structure: The Legacy of PPB

There are few prerogatives of Congress that are more highly cherished than its "power of the purse." Just as important as its power to direct federal budgetary resources to particular agencies and programs is its power to see that the appropriations are actually spent in accordance with Congress' intent. The General Accounting Office (GAO) was established as an investigative arm of the Congress largely to monitor spending and other financial activities of federal agencies and assure that they are in compliance with federal appropriations law.[1] Yet the Forest Service's reporting to Congress of the apportionment of its expenditures, as well as the outputs associated with those expenditures, has become so inaccurate that it is no longer a reliable representation of actual spending and is of questionable value in Congress' control of appropriations and direction of the activities of the Forest Service.

The extent of the inaccuracies in the Forest Service's financial reporting to Congress has only recently become known to the Congress, and perhaps even to the leadership of the Forest Service itself. The weaknesses of the Forest Service financial reporting system began to surface in the context of a House Interior Appropriations Subcommittee investigation into the sale of national forest timber at a net financial loss to the federal government. Pressed by the subcommittee to provide information on the costs and revenues associated with the timber sale program, the chief of the Forest Service reported that the current accounting system simply was not capable of providing detailed information on the cost of the program.[2] A subcommittee-requested investigation by GAO into the question of "below-cost timber sales" on the national forests reported that, using the information available from the Forest Service's current accounting system, it was not even possible for the agency itself to determine whether below-cost timber sales were taking place and, if so, to what extent.[3] The Forest Service was ordered to improve its accounting system, and soon after began testing a revised accounting system on several national forests.

It was in an analysis of the Forest Service's application of the proposed accounting improvements that GAO reported finding much more fundamental problems than could be addressed by the proposed improvements. Among their findings was that the existing system was simply incapable of providing accurate cost information because the direct time charges by field personnel, by far the largest area of expenditures, were themselves full of inaccuracies. In an attempt to show expenditures to be consistent with appropriations for each program, noted GAO, personnel were not reporting their time as it had actually been worked, but as the time had been allocated in their work *plans*.[4] The expenditures being reported to Congress bore no direct relationship to how the funds had actually been spent. Further, the expenditures being reported bore no direct relationship to the various resource outputs being reported as target accomplishments. Table 8.1 clearly illustrates these findings for the five case study national forests. Some of the most fundamental financial information an accounting system should provide for managers, such as unit costs of production for various resources both over time and between the different national forests, was either missing or erroneous under the current system. The financial accounting system then in use, the Program Accounting and Management Attainment Reporting System, was not only of little use to the Congress in exercising its oversight responsibilities over the Forest Service, it was essentially useless to as a management tool for the Forest Service itself.

The decline in the usefulness of the financial reporting system for this purpose did not occur overnight. It is a function of the longer-term evolution of the management and control mechanisms for the Forest Service as an organization. This evolution is inextricably tied to the evolution of the Forest Service's internal budget development process from one highly dependent on accurate and reliable unit cost information to one in which unit cost information plays no direct role. Up through the late 1960s, the agency budget request was developed almost entirely in the Washington office. The field units each year projected what their resource outputs, primarily timber sale volume, would be, and cost estimates were developed by the Washington office through a complex process of continually monitoring cost reporting, and periodically auditing and observing the activities on dozens of the more than 650 ranger districts. A brief description of the process in the mid-1960s is as follows:

> The Washington office maintains a standardized list of specific jobs--that is, discrete operations characterized by countable work output at specified standards of quality (perfection and intensity)--that must be done to perform all the functions in national forest administration. It also keeps on hand records of the "converting factors"--the amount of time (down to the smallest fraction of an hour) needed to perform one unit of work on each job. Finally,

Table 8.1
Actual Budget Trends and Plan Implementation Budget Requirements for Case Study National Forests

CASE STUDY FOREST AND PROGRAM	FY 83	FY 84	FY 85	FY 86	FY 87	FINAL ALLOC FY 88	INITIAL ALLOC FY 89	FOREST PLAN FY 89	PERCENT OF FY 89 FOREST PLAN
FOREST 1		[Thousands of nominal dollars]							
TIMBER	986	1,117	1,082	1,134	1,131	1,011	1,091	1,384	79%
RECREATION	509	494	680	659	704	927	753	936	80%
FISH & WILDLIFE	170	170	182	160	396	512	270	739	37%
RANGE	73	64	111	63	89	99	98	147	67%
SOIL & WATER	86	101	110	109	122	148	137	148	93%
MINERALS	30	12	47	53	58	45	46	49	94%
PROTECTION	40	53	47	42	44	48	47	58	81%
FACILITIES	208	231	211	317	242	274	239	396	60%
LANDS	236	280	312	310	410	338	336	508	66%
FOREST 2									
TIMBER	986	1,117	1,082	1,134	1,131	1,011	1,091	1,384	79%
RECREATION	509	494	680	659	704	927	753	936	80%
FISH & WILDLIFE	170	170	182	160	396	512	270	739	37%
RANGE	73	64	111	63	89	99	98	147	67%
SOIL & WATER	86	101	110	109	122	148	137	148	93%
MINERALS	30	12	47	53	58	45	46	49	94%
PROTECTION	40	53	47	42	44	48	47	58	81%
FACILITIES	208	231	211	317	242	274	239	396	60%
LANDS	236	280	312	310	410	338	336	508	66%
FOREST 3									
TIMBER	11,127	8,332	5,685	5,719	8,233	8,217	9,397	9,725	97%
RECREATION	888	823	707	855	1,367	1,586	1,564	1,863	84%
FISH & WILDLIFE	225	315	317	316	425	466	416	729	57%
RANGE	53	49	52	37	50	52	40	80	50%
SOIL & WATER	209	232	196	229	273	263	286	420	68%
MINERALS	145	185	203	170	172	134	214	365	59%
PROTECTION	912	851	838	819	847	941	912	1,358	67%
FACILITIES	1,157	1,170	928	820	941	1,001	1,114	1,261	88%
LANDS	406	374	376	301	267	460	519	720	72%
FOREST 4									
TIMBER	3,513	5,868	4,536	3,839	3,278	2,922	2,741	5,338	51%
RECREATION	491	489	698	545	785	987	1,086	1,569	69%
FISH & WILDLIFE	151	147	136	229	260	407	553	966	57%
RANGE	640	491	650	446	544	578	512	919	56%
SOIL & WATER	322	316	343	377	324	415	292	618	47%
MINERALS	57	65	61	52	79	76	101	123	82%
PROTECTION	609	726	729	689	755	841	823	1,127	73%
FACILITIES	449	938	980	965	755	758	704	1,213	58%
LANDS	118	181	132	153	168	193	188	266	71%
FOREST 5									
TIMBER	14,098	14,659	13,886	12,705	14,789	15,284	16,811	18,551	91%
RECREATION	1,016	1,333	1,112	1,253	1,982	2,888	2,693	4,051	66%
FISH & WILDLIFE	546	460	702	628	1,022	1,438	1,338	1,745	77%
SOIL & WATER	310	104	5	249	311	397	468	694	67%
MINERALS	76	75	92	105	95	110	127	131	97%
PROTECTION	1,231	967	903	915	1,203	1,003	903	964	94%
FACILITIES	1,715	1,641	1,725	1,536	1,795	2,096	2,119	2,261	94%
LANDS	691	606	534	418	428	491	599	657	91%

> it has figures on the number of units of work--the volumes of business--for each job. Multiplying the volumes of business by unit time allowances yields the number of man-hours required to perform all the functions involved in managing the national forests. The man-hours are converted into dollar sums. . . . Thus, the estimates are formulated in terms of physical performance. The performance information comes in from the field; that is why the Washington office is able to turn out the agency budget without getting dollar estimates from the subdivisions.[5]

The time allowances for a "unit of work" varied from region to region and required extensive and continuing field studies. Teams from each regional office selected sample ranger districts at random and tracked the amount of time expended by personnel to accomplish various types of work. Areas of the country where the allowances were fairly uniform were grouped into zones without regard to the boundaries of the administrative regions. In fact, different zones were defined depending on the resource activity in question--for recreation the entire country was one zone, for timber management it was divided into five zones, and for fire suppression into seven zones--and any given national forest may find itself in different zones for different functions. In addition to this observation and sampling, the organization placed a very strong emphasis on detailed, accurate accounting of the time spent by each employee in each aspect of his or her duties. Until the mid-1960s, each district ranger was required to maintain a diary of his activities, and the review of this diary was an important part of his annual performance evaluation. Even after the diaries were no longer required, daily logs of every individual's activities ("the old 1300 dash 8s") were produced and kept on file for several years. The detail of these reports and the organizational emphasis on precision created a solid basis for the compilation of accurate unit costs as the basis for Washington office budget development.

Such a system obviously required tremendous amounts of time and effort, and diverted a great deal of personnel time from actual management activities in the field to administrative paperwork in the office. Although this system functioned quite well with only minor changes over the first six decades of the Forest Service's history, the rapid expansion of the agency's management activities during the 1960s soon made these detailed reporting and inspection requirements so burdensome that they ceased to function effectively. When the Planning, Programing and Budgeting (PPB) system came into wide use in the federal government in the late 1960s, it was enthusiastically received by the Forest Service. Here was an opportunity to revamp their obsolete and increasingly unworkable system for planning and budgeting, and PPB seemed particularly well suited to the Forest Service management situation. Three basic concepts underlie PPB:

(1) the definition of the organization's objectives in specific, operational terms, and the generation of alternative multi-year programs for meeting those objectives;

(2) an annual cycle that involves the generation and modification of program proposals, evaluation of the proposals using various analytical tools, and selection of appropriation proposals for funding; and

(3) a management reporting system that monitors accomplishment of the funded programs, as well as providing input for program generation.[6]

By the very nature of its responsibilities in managing millions of acres of renewable resources on a long-term sustained yield basis, the Forest Service already had a great deal of practical experience with multi-year planning based on well defined, operational objectives. Because of this prior experience, and because its outputs were more easily defined than those of most other federal agencies, the Forest Service had a great deal more success with PPB than most of the other agencies that experimented with it.[7] President Johnson had begun the federal experiment with PPB by executive order in 1965, and by 1970 it had been declared moribund by most federal agencies.[8] However, early experiments with PPB by the Forest Service were declared a success and, in 1970, Chief John McGuire made the decision to officially adopt the system. The prime objective was "the establishment of an ongoing process that would link the field units (regions, forests, and districts) to the Washington office through a unified program planning package."[9]

A subtle but important feature of PPB is its strong centralizing bias. By design, it provides the means by which an agency's philosophies, policies and long-term goals can be translated into definable, achievable shorter term objectives, and the agency's scarce budgetary resources can be allocated among alternative objectives.[10] To a much greater degree than in the past, national forest output objectives were decided upon in Washington and disaggregated through the administrative levels of the Forest Service to the field. This came in the form of budget development instructions developed by the Washington office, with guidance from OMB, which required the preparation of five alternative program proposals corresponding to different levels of funding and activity. At the same time that the Washington office was becoming more involved in the setting of output objectives, the field was by necessity becoming more involved with determining the funding levels that would be associated with the different levels of resource outputs. The Washington office's elaborate system of maintaining precise unit cost estimates for every activity in every zone of the National Forest System fell by the wayside and there was an increasing reliance on the field to simply provide their own estimates of the funding required to achieve different levels of resource outputs. A natural tendency on the part of field officials to begin to inflate these estimates was damped off by the fact that, despite the advent of PPB, the budget and appropriations process was still fundamentally an incremental process, and neither OMB nor the Congress was ready to provide major increases in Forest Service funding. This was perhaps the greatest frustration of field officials with the new Forest Service budgeting system, with many regarding it as a "Washington-oriented paper pushing exercise" that produced no real change. The complaints from Forest Service field managers were not unlike those of officials in other federal agencies who had not found PPB living up to its promise and potential.

> These managers base their case on the fact that the budget which they are allocated may be significantly different from the program proposals they submitted for that year. Typically, they argue that their budgets are very close to being extensions of historical trends, with exceptions in a few areas each year. PPB, they say, is not effective as a means of changing directions in funding and is, therefore, not worth the manpower required to implement it.[11]

Nonetheless, the direct link between the Forest Service's accounting system and its internal budget development system had been broken, setting in motion a process that allowed the unit costs implicit in the annual statements of expenditures and outputs to drift further and further afield. When unit costs were an essential ingredient in the agency's budget development process, there was a strong organizational emphasis on the completeness and scrupulous accuracy of the accounting data, especially regarding the time charged to different work activities as it was used to determine the expenditures necessary to carry out those activities. There were powerful organizational incentives, tied to staff evaluations and staff advancement, for exacting performance in this area, and powerful organizational sanctions, including demotion or dismissal, for poor performance. Under the PPB-based system, the Forest Service as an organization no longer relied on the unit costs implied by reported expenditures and outputs, but on direct estimates of program costs flowing up from the field.

The financial reporting system developed subsequent to the Forest Service adoption of PPB, the Program Accounting and Management Attainment Reporting System, is almost entirely a device to permit the Forest Service to show the Congress that its appropriations to the agency were spent as directed in the appropriations bill and committee reports. The emphasis placed on accurate accounting by the upper ranks of the organization, while not willfully or explicitly reduced, was nonetheless diminished by the simple fact that it was no longer a critical need of the organization itself, but a perfunctory exercise performed for the benefit of Congress (which, after all, did not have sufficient regard for their professional judgment to fund their activities as requested through the budget development process).

Thus, Forest Service field officials have received a powerful complex of unwritten and unspoken signals indicating that detail, precision and accuracy in accounting is, in effect, of secondary importance to their organization. Because accurate unit cost data is no longer necessary for the agency to develop its budgets, there is no particular organizational incentive for field personnel to be scrupulously careful in their accounting. And because the flexible approach to accounting that has been taken has not caused any particular problems for the agency (until recently), there have been few if any organizational sanctions for *not* striving for precision. On the other hand, field officials are under constant pressure to meet timber and other resource output targets that have been assigned them and, at the end of the fiscal year, to show each one of their myriad fund accounts to be "zeroed out." They know that these considerations will be the two greatest factors in their next performance evaluation and thus their chances for future career advancement. These are the considerations they *must* pay attention to. To the extent that they must compromise responsibilities for which they will be closely evaluated in order to attend to responsibilities for which there are few if any organizational incentives, this second set of responsibilities will be neglected. The cardinal rule for a ranger is to get the job done and not allow issues to develop beyond the ranger district. As seen on the national forests studied, field managers and their budget staffs make every effort to keep expenditures (i.e., the recorded allocation of staff time) consistent with their fund allocations on a program by program basis. Work is charged to individual management codes without a rigorous regard for the work being done, however. If the management code does not include certain activities that nonetheless must be accomplished, the work is done and its costs charged to one or more of the operative set of management codes. The disincentives associated with not containing issues at the lowest organizational level are far stronger than the incentives to account strictly by the book and let issues get out of hand.

In terms of the day-to-day management of thousands of acres of national forest land, wherein timber management, recreation use, fish and wildlife habitat management, livestock grazing and watershed protection are physically and geographically integrated, the accounting requirements on the average district ranger are indeed daunting. Currently, Forest Service field managers receive funds for national forest management activities in as many as 49 different accounts. Small portions of as many as a dozen of these accounts may be integrated into a single project, of which there may be 50 or more in a single year. At the close of the fiscal year on September 30, the district ranger must be able to show that each of these 49 or so accounts has been spent to within a very small percentage of what was allocated, often amounting to less than a $100 deviation on an account of tens of thousands of dollars. Organizational norms present strong disincentives, including but not limited to unfavorable performance evaluations, for showing any significant departure above *or* below the allocated amount. Departures above the allocated amount incur the additional disincentive of a possible criminal violation under the Anti-Deficiency Act.

The Forest Service prepares its budgets and accounts for its expenditures in greater detail than is actually required by appropriations law, guided by specifications in the appropriations committee reports that have gotten more and more detailed over time. Technically, there are only five different appropriations to the Forest Service, one of which is "National Forest System," and expenditures must match appropriations at this level of aggregation. Congressional direction through the appropriations committee reports, however, is to maintain both budget presentation and accounting to the detail found in the House or Senate appropriations committee report, whichever is the most detailed. If for some reason the Forest Service wishes to change the detail shown in the budget presentation, the agency must receive prior approval from both the House and Senate appropriations committees. Currently, the five appropriations are broken down into 49 separate budget line items, each of which becomes an account to be tracked by the accounting system. As indicated above, if a manager senses that expenditures will differ from what was allocated to his or her unit under a particular budget line item, a request for reprogramming may be submitted to move funds from an underrun account to an overrun account. As discussed earlier, the extremely limited authority of the Forest Service to reprogram ($250,000 or 10%, whichever is less, for each budget line item, *nationwide*), and the problem of knowing in time whether or not a reprogramming request has been approved make reprogramming of little practical use to field managers in correcting mismatches between program funding and the activities needed to achieve the assigned output targets. Given the shortcomings of the formal flexibility granted the field officials for dealing with funding mismatches, an informal system has developed. Unfortunately, this informal system has produced years of increasingly inaccurate records, raising serious questions about the accountability of the Forest Service and the ability of Congress to effectively oversee its activities.

NOTES

1. Comptroller General of the United States, *Principles of Federal Appropriation Law* (Washington, D.C.: General Accounting Office, 1986). The statutory requirements for the legal expenditure of appropriations are found in 31 U.S.C. 1301, 1341 and 1502.

2. Testimony of R. Max Peterson, in: U.S. Congress, House of Representatives, Committee on Appropriations, *Hearings on Appropriations for the Department of Interior and Related Agencies for Fiscal Year 1985.*

3. U.S. Comptroller General, *Congress Needs Better Information on Forest Service's Below-Cost Timber Sales*, GAO-RCED-84-96, 6/28/84 (Washington, D.C.: General Accounting Office, 1984).

4. U.S. Comptroller General, *Timber Sale Accounting: Analysis of Forest Service's Proposed Timber Program Information Reporting System*, GAO/AFMD-86-42, 4/4/86 (Washington, D.C.: General Accounting Office, 1986); U.S. Comptroller General, *Timber Sale Accounting: Analysis of the Kootenai National Forest*, GAO/AFMD-86-62BR, 6/6/86 (Washington, D.C.: General Accounting Office, 1986);

5. Kaufman, Herbert, *The Forest Ranger: A Study in Administrative Behavior* (Baltimore: Johns Hopkins University Press for Resources for the Future, 1967), pp. 114-118.

6. Novick, David, *Program Budgeting* (Cambridge, MA: Harvard University Press, 1967).

7. Zimmerman, Kristen E., *Capital Budgeting and Its Pertinence to the Management of the U.S. Forest Service*, Unpublished Research Paper, Harvard Graduate School of Business Administration, January, 1976, pp. 13-16.

8. Harper, Edwin L., Kramer, Fred A., and Rouse, Andrew M., "Implementation and the Use of PPB in Sixteen Federal Agencies," *Public Administration Review*, Vol. 29, No. 6 (Nov/Dec 1969), pp. 631-632. See also: Harlow, Robert L., "The Decline and Possible Fall of PPB," *Public Finance Quarterly*, (January 1972), pp. 85-115; Schick, Allen, "A Death in the Bureaucracy: The Demise of Federal PPB," *Public Administration Review*, Vol. 33, No. 2 (March/April 1973), pp. 146-156.

9. Gremillion, Lee L., McKenney, James L., and Pyburn, Philip J., "Program Planning in the National Forest System," *Public Administration Review*, Vol. 40, No. 3 (May/June 1980), pp. 226-230.

10. Wildavsky, *The Politics of the Budgetary Process*, p. 189.

11. Gremillion, McKenney, and Pyburn, "Program Planning and Budgeting in the National Forest System," p. 229.

9

Revision of the Forest Service Budget Structure: Problems and Potential

The response of Forest Service leaders to congressional concern over the reliability of its accounting information has been partly to send sternly worded directives down through the ranks advising field officials to shape up and account for their activities on a "charge as worked" basis.[1] This hardly changes the fundamental situation that gave rise to the problem, however, so a second response of top agency officials has been to approach Congress about loosening their appropriations control to something approaching the current legal requirements and concentrating instead on setting desired output levels for the full range of national forest resources.[2] The objective, in effect, is to encourage Congress to at least implicitly recognize the need for an integrated approach to funding as well as planning national forest resource management activities. Effective functional integration in the field is heavily dependent upon effective functional integration in Washington, not just in the minds of Forest Service leaders but in the minds of other key governmental organizations that strongly influence the agency's programs.[3] The Forest Service itself has taken a number of administrative measures over the years to encourage functional integration within its own ranks, such as the shifting of budget development and allocation authority from functional staff to the line officers. In preparing its annual budget proposal, however, the Forest Service is required to take budget information flowing up from the field, which is based on integrated local resource management plans, and reconfigure it to conform on paper to the dozens of budget line items specified by Congress and OMB. Much of the integration is lost during this translation and in the subsequent modification of individual budget line items in relative isolation from one another.

The main thrust of the Forest Service proposal, generally referred to as "end-results budgeting," would collapse the current 49 budget line items into only nine budget line items (see Figure 9.1). Funding for all of the different resource programs on the national forests would be lumped under one budget line item so that funds could be allocated to the field on a more integrated basis, and with no

Figure 9.1
Proposed Budget Structure Under "End Results" Budgeting
(National Forest System Only)

EXISTING BUDGET LINE ITEMS (43)	PROPOSED BUDGET LINE ITEMS (6)	RESOURCE OUTPUT MEASURES
Cultural Resources		# Acres
Fighting forest fires		
Wildlife and fisheries administration		
Wildlife habitat improvement		# Acres
Inland fish habitat improvement		
Endangered and threatened species habitat improvement		
Anadromous fish		# Structures
Range vegetation management		# Acres
Range structure improvements		
Wild horse and burro management		
Range forage improvements		
Noxious weed control		# Acres
Soil, water and air administration		
Soil and water resource improvements		# Acres
Soil and water resource inventory		
General administration		
	} Operations and maintenance	
Locatable minerals		
Leasable minerals		# Leases
Common variety minerals		
Geology		
Land exchange and adjustment		# Acres
Classification status		
Special uses		
Geometronics		
Land line location		# Miles
Presuppression of forest fires		
Forest fires fuels management		# Acres
Cooperative law enforcement		# Patrol units
Road maintenance		# Miles
Trail maintenance		# Miles
Harvest adminstration		
Timber resource planning		
Silvicultural examination		
Sales preparation		MM Board Feet
Recreation management		M Persons At One Time
Wilderness management		# Acres
Facilities construction	} Facilities construction	
Road construction	Road construction Road reconstruction	
Trail construction and reconstruction	} Trail construction and reconstruction	
Reforestation Timber stand improvements Nursery and tree improvements	} Reforestation and timber stand improvement	
Land acquisition	} Land acquisition	

institutional obstacles impeding the movement of funds from one resource program to another. What control the Congress would be giving up by no longer directing expenditures on individual resource programs would be compensated for, under the proposal, by the setting of approximately seventeen well-defined resource output targets in each year's agency appropriations. Detailed quarterly progress reports would keep the Congress apprised of the agency's success in achieving the assigned targets. Interestingly, *accounting* would still be on the current basis of 49 or more individual accounts, the difference being that these would be for reporting purposes only--field officials would not be constrained to "zero out" these individual accounts at the end of the fiscal year, only the nine recast budget line items. Under such a system, the Forest Service contends, there would be no need for field officials to fudge the charging of their time and the reporting of their expenditures; the resulting accounting records would automatically become more accurate, producing valid unit cost information that would be a valuable tool for both congressional oversight and Forest Service managerial control.

Since 1984, the Forest Service has been testing elements of the proposed budgeting system on three of the national forests in three different administrative regions and claims considerable success. The test forests were chosen on the basis of their relatively small annual budgets so that, should actual expenditures depart dramatically from those in the past, the overall National Forest System could absorb the differences and still be within the existing congressional appropriations control requirements. The differences were found to be quite manageable, however, and in 1986 the test was expanded to include all of the Eastern Region (the Lake States, Midwest and New England). As might have been expected, the general trend was that the implicit unit costs of producing timber and minerals began to decline and the unit costs for many of the non-commodity programs was either unchanged or tended upwards. During FY 1987, one of the test forests saw five percent of its budget redirected primarily from fire protection, timber and minerals programs to the range, recreation and fish and wildlife programs. In the Eastern Region, the cost of all resource programs across the board were reduced by about two percent, reportedly due to increased efficiency due to improved integration of management activities. In this pilot program, as in the current proposal, the proceeds from an underrun in one program may be retained on that national forest and redirected to other programs where funding is less than adequate. That is, whatever "savings" was achieved was not simply returned to the general Treasury. OMB has, on principle, opposed such provisions in the past for any federal agency, but the Forest Service regards this as a key element to the success of the overall proposed system.

Asked by the appropriations committees to evaluate the Forest Service "end-results budgeting" proposal, GAO was favorably impressed and supported the Forest Service's tentative plans to begin implementation of the new budgeting system nationwide at the start of fiscal year 1989.[4] In its report to the House Interior Appropriations Subcommittee, GAO noted a number of significant advantages of the proposed system for both the Forest Service and the Congress itself. GAO investigators had spent several months working with Forest Service officials in Washington and in several of the field offices and recognized the value of less finely differentiated appropriations to more efficiently carrying out the legal mandate for multiple-use management of the national forests.

> By measuring program performance through output targets, the Forest Service, the Department of Agriculture, and the Congress would be able to view results of their operations in relation to planned work. In addition, by allowing increased flexibility to redirect certain funds to other priorities, the Forest Service expects to fund work where priorities dictate rather than where funds were originally planned. These factors could potentially heighten managers' achievement of program results.[5]

GAO confirmed the Forest Service's claim of increased overall cost efficiency on the test forests through better program integration. An important element in these decreased program costs could well be in the time spent by managers actually directing field operations rather than fine-tuning their accounting records. GAO indicated in its report, "the amount of additional time that managers said could now be spent in the field rather than on budget and financial matters ranged from 10 to 20 days a year," and noted that such "reductions in administrative time at a ranger district could mean important increases in resource management productivity of the Forest Service as a whole."[6]

Ironically, by adopting a budgeting system that ostensibly diminishes the direct appropriations control that Congress has over the Forest Service, congressional control and oversight capabilities, and thus the Forest Service's accountability, may in reality be enhanced. GAO recalled their earlier finding that district rangers' "normal reaction to being held responsible for numerous accounts at year-end led to a system whereby the amount charged to an account was often based on the amount budgeted for that account rather than the actual work chargeable to the account," but it confirmed Forest Service reports that this practice was almost entirely eliminated on the forests testing the proposed end-results budgeting process. This led GAO to report to the appropriations committee that:

> Although less financial information will be displayed in the proposed budget structure, the level of detailed information on actual expenditures will be the same as currently provided through the Forest Service's accounting system.

> Moreover, the Congress could potentially receive, under the end-results budget proposal, budget information that is more accurate than the information it now receives.[7]

Though stated as diplomatically as possible, GAO was essentially telling Congress that the accounting information they were receiving under the present system was telling them little about what was actually being done out on the national forests and that, by relinquishing what was in fact not much more than a pretense of oversight capability, they could gain a significant measure of real control over the agency's activities.

Despite the generally glowing reports and positive recommendations of GAO, there are some problems that remain to be solved before end-results budgeting can measure up to its claims. The first is the "softness" that exists in many of the current output measurements for all but the commodity resources such as timber. As pointed out in the discussion of FY 1986 budget execution on the five case study forests, many of the non-commodity resource output measurements used under the current Management Attainment Reporting system are extremely fluid and based much more on subjective judgment than on any physical measurement. Should such output measures continue to be used under an end-results budgeting system with only a few broad categories of appropriations control, there is a particular danger that the non-commodity programs would receive less attention than they do now rather than more. With the continuation of current trends in federal budgeting, there is a high probability that overall funding for the Forest Service will continue to decline in real terms. Also, for several years now, the appropriations committees have asked the Forest Service to offer more timber sales than the agency proposed, and with less funding than the agency requested. If the Forest Service continues to find itself caught in this double bind, there will be a strong tendency for field officials to continue to place the highest priority on accomplishing the "hard" targets assigned to them, knowing that the others can be fudged if necessary. And once the institutional barriers to shifting funding among programs have been dropped, there may well be a net flow of funds away from rather than toward the non-commodity resource programs. In the pilot study conducted by the Forest Service, the net flow of funding on one national forest was away from timber and toward recreation; on another national forest the net flow was the reverse.[8]

If measurement and reporting of the accomplishment of resource output targets is to play an even greater role in congressional oversight of national forest management activities, then their validity and usefulness must be considerably improved. One resource output currently a part of the Management Attainment Reporting system that has been singled out by the House Interior Appropriations Subcommittee as particularly useless is the number of acres of wilderness managed. On any national forest, this number is not a function of any given level of effort on the part of managers, but of the geographic area so designated by Congress. Even a major increase or decrease in funding, and in Forest Service management efforts, would have no impact on this number. As observed by GAO, such figures "do not directly relate to the amount of effort needed to monitor or maintain these acres." In effect, they measure nothing and mean nothing. The introduction of some indicator of *quality* of management, rather than simply quantity of outputs may help in some instances. In developing their resource management plans, many national forests have zoned wilderness areas according to the level of environmental quality or "wilderness characteristics" to be found there. One way of improving the measurement of wilderness management "outputs" might be to report the number of acres being managed at or above one of several clearly defined levels of environmental quality, as objectively determined through the use of a set of physical, biological and social indicators, and different standards for each level.[9] Similarly, the number of miles of trail maintained means little, but the number of miles maintained at each of several successive levels of quality conveys information that is a great deal more useful for management evaluation and oversight.

Even the output measurements for timber could be made much more useful for informing and guiding the decisions of Congress. Reducing funding and raising assigned output levels may seem like an effective, if crude, way of trimming the fat out of the timber budget--especially when the timber targets have somehow always been met. If there are costs of such an approach in terms of the *quality* of timber management on the national forests, then Congress needs to know about them. Such tight budget constraints can result in the violation of environmental quality standards that the Congress, through such laws as the National Forest Management Act, has declared to be important. Funding not commensurate with congressional timber sale volume requirements can force an overconcentration of timber sales within a small area and increase soil erosion and water quality degradation. Forcing field officials to opt for the least costly methods of sale preparation and administration may result in excessive reliance on clearcutting rather than permitting the development of silvicultural systems wherein timber volume removals are stretched out over a longer period of time, also an explicit objective of NFMA. Perhaps Congress would be more usefully informed if, in addition to simple reports of timber volume sold, they also had information that could convey the quality of the operations--number of acres to be harvested under systems other than clearcutting, or number of tons of soil loss due to the disturbances of roadbuilding and timber harvesting.

In redesigning the system of national forest output measurements, both Congress and the Forest Service must cooperate to assure that the output to be measured is consistent with the goal actually being sought. Those that are not will encourage a tendency toward a sort of "goal displacement," the means being treated as goals in themselves, that can be both misleading and counterproductive.[10] Static output measurements may be no more indicative of progress toward long-term resource management goals than body counts, tons of bombs dropped or number of sorties flown are of progress toward winning a war. Changes in resource outputs or resource conditions over time can be a useful way of displaying trends over time in order to induce corrective actions before unacceptable levels are actually reached. A recent Forest Service annual report contained a chart showing Congress the impact of cutbacks in recreation funding in terms of an eight-year decline in the number of facilities maintained "to standard" and the corresponding rise in the millions of dollars worth of deferred maintenance. Such a display conveys much more information than a single year's point estimate of these two indicators.[11]

A second problem to be overcome before end-results budgeting can work is a mutual lack of trust between the Congress and the Forest Service, especially the field officials of the Forest Service. For the Congress, there is a certain amount of indignation over having been for years provided inaccurate accounting information by the Forest Service. This has prompted the appropriations committees to require the Forest Service to improve the accuracy of its accounting as a *precondition* for considering a less detailed budget structure. The appropriations committees have provided flexibility for dealing with funding mismatches through the reprogramming authority granted to the chief and have expressed their willingness to promptly consider any reprogramming needs that exceed that limited authority. In their view, the Forest Service should make full use of the discretion it has already been granted before requesting what amounts to a substantial broadening of that spending discretion. Forest Service field officials, on the other hand, see themselves in somewhat of a "prisoner's dilemma." For many years, they have seen their budget proposals warped by the appropriations process, making it difficult for them to fulfill their perceived responsibilities for multiple-use forest resource management on the ground. They know that, for whatever reason, the timber and minerals programs are the only programs that tend to get funded at or above the level requested by the agency, with funding for most other programs being significantly reduced from the agency request. Should they act unilaterally to provide the Congress with more accurate program costs, they could lose what political leverage they now have and find themselves in a worse position than when they began. An overriding concern encountered among field officials in the case study forests is that, by their indicating to Congress that unit costs for timber production are lower than have been reported in the past and unit costs for the non-commodity programs higher, the appropriations committees may indeed cut funding for timber (while maintaining the same output requirements) but not provide the corresponding increases for the other programs. There is, after all, a considerable probability that Congress will not give up its direct control on appropriations just because the Forest Service has asked them to, and field officials, in spite of earnest letters from the chief, are not enthusiastic about putting themselves out on a limb with Congress holding the saw.

Clearly, if neither party is willing to make the first move there can be no progress. Through the complexity of its tasks, its large numbers of people with highly specialized expertise and its control over the flow of information, the Forest Service has considerable control over its own destiny; Congress' power over the agency is therefore far from absolute. Effective and lasting policy change in such an environment of shared power calls for a gradual *rapprochement* between Congress and the Forest Service field cadre, a process based upon incremental *quid pro quo* and a progressive building of mutual trust. The chief of the Forest Service plays a central role in this process, not unlike the role often played by the president in relations between the federal bureaucracy and the Congress. The chief's power as a leader is the power to persuade.[12] To lead he or she must have both the dispensation of the Congress and the confidence of the 35,000 professionals at the operational levels of the organization. Understanding that key elements of power reside with each side, the chief must in effect play the role of a mediator, encouraging a gesture of trust from one side and a good faith response from the other. In encouraging the Congress to have enough faith in the Forest Service to begin implementing the end-results budgeting proposal, Chief Dale Robertson has first sought to gain the personal trust of the appropriations committees:

> One assurance you have is my commitment to you. I'm directing Forest Service field people to charge as worked, not charge as planned. And I believe Forest Service people will carry out that commitment. . . . That may mean we exceed the reprogramming requirements of this committee, when you add it all up for 650 Ranger Districts after the fact. But I want accurate unit costs. I think you want accurate unit costs and we need to make sure we get that.[13]

Following the House Interior Appropriations Subcommittee hearing at which this testimony was presented, Robertson took the unusual step of sending a very frank letter to every regional forester, forest supervisor and district ranger in the Forest Service, reprinting this excerpt from his testimony and openly seeking the support of the agency's rank and file:

> I want to urge each and every one of you to help the Forest Service honor that commitment. Congress may approve some of our end-results budgeting proposal for FY

> 1989, but I know we will not be able to gain further approval until they are assured we are providing good unit costs. . . . My intent here is to ask each of you to help us close out this fiscal year with a strong commitment to the charge-as-worked philosophy.[14]

To the Congress, the chief portrayed himself as willing to personally assume the risk that Forest Service officials across 650 ranger districts would make the leap of faith and begin accounting for their activities more accurately. To accomplish this, the chief's letter essentially provided assurances to field managers that he would also personally assume the risk that individual resource programs might incur overruns when aggregated to the national level.

Thus far, this seems to be an effective approach to gaining the trust and cooperation of both the Congress and Forest Service field officials. Congress is certainly not ready to relinquish direct control over appropriations until it is assured that the establishment and monitoring of resource outputs provides a comparable degree of control over Forest Service activities. Nor can we expect Congress to begin considering this type of budgeting for federal agencies generally. Despite the difficulties involved in establishing clearly defined output measurements for some of the non-commodity resources, the goods and services from the national forests are still easier for members of Congress to relate to than those of most public sector organizations. How manageable these problems turn out to be, and how well the Forest Service can bring along both the Congress and its own field managers, will have major implications for the agency's success in implementing its program planning efforts, particularly the forest plans prepared under the National Forest Management Act.

NOTES

1. USDA Forest Service, Washington Office, "Fiscal Integrity in Financial Reports," Memorandum from Deputy Chief, Administration, to Regional Foresters, January 28, 1988.

2. USDA Forest Service, *Forest Service End-Results Budgeting Proposal*, Internal Working Paper, March 23, 1988.

3. Fesler, James W., "The Basic Theoretical Question: How to Relate Area and Function," in: Grosenick, Leigh E. (ed.), *The Administration of the New Federalism: Objectives and Issues* (Washington, D.C.: American Society for Public Administration, 1973), p. 10.

4. U.S. Comptroller General, *Forest Service: Evaluation of "End-Results" Budgeting Test*, Report to the Subcommittee on Interior and Related Agencies, Committee on Appropriations, House of Represen-

tatives, GAO/AFMD-88-45 (Washington, D.C.: General Accounting Office, 1988).

5. Ibid., p. 4.

6. Ibid.

7. Ibid., p. 6.

8. Ibid., p. 25.

9. Such a measurement system could be readily devised from the already well developed wilderness management concept of "limits of acceptable change," wherein a clearly defined set of physical, biological and social indicators are established for each of several zones within a wilderness area, with different thresholds of environmental quality set for each zone. See: Stokes, Gerald L., "Involving the Public in Wilderness Management Decision Making: The Bob Marshall Wilderness Complex Case Study," in: *Proceedings of the Annual Meeting of the Society of American Foresters* (Bethesda, MD: Society of American Foresters, 1988).

10. See: Merton, Robert K., "Bureaucratic Structure and Personality," in: Merton, R. K., Gray, Ailsa P., Hockey, Barbara, and Selvin, Hanon (eds.), *Reader in Bureaucracy* (Glencoe, IL: Free Press, 1952), pp. 361-371.

11. USDA Forest Service, *Report of the Forest Service, Fiscal Year 1985* (Washington, D.C.: USDA Forest Service, 1986), p. 24.

12. Neustadt, *Presidential Power*, pp. 101-114.

13. Testimony of F. Dale Robertson, Chief, Forest Service, in: U.S. Congress, House of Representatives, Committee on Appropriations, *Hearings on Appropriations for the Department of Interior and Related Agencies for Fiscal Year 1989* (Washington, D.C.: Government Printing Office, 1988), pp. 826-831.

14. Letter from F. Dale Robertson, Chief, Forest Service to Regional Foresters, Station Directors, Area Director and Washington Office staff directors, May 23, 1988.

10

Summary and Conclusions

Summary of Findings

At this writing, resource management plans required by the National Forest Management Act have been completed for nearly all of the 156 national forests administered by the Forest Service. In spite of its long-standing use of management plans to assess national forest resources and guide their protection and utilization, the Forest Service has never in its history invested as much effort and expense in as complex and comprehensive a planning effort as it has made to produce these forest plans. The agency is now faced with the equally unprecedented task of translating the management policies and responsibilities represented in these plans into the day-to-day operations of managing the national forests. In an effort to gain some insight into what can be expected during the implementation phase of national forest planning, it has been useful to look at the decade or more of the Forest Service's implementation of the Resources Planning Act, the progenitor and foundation for NFMA. Forest planning and RPA planning continue to be intertwined, and both must be integrated effectively with the agency's annual budget development process to comprise the entire framework within which operational decision making takes place on the national forests.

Budget Development

Each of the RPA Programs to date have called for major real-term funding increases in all national forest resource programs. They have also called for a gradual but steady shift in program emphasis on the national forests. The RPA Assessments had identified a rapidly growing demand for the entire range of forest resources in the United States, especially for many of the non-commodity resources. The Programs developed for the national forests attempted to address the full range of needs, but recognizing the greater suitability of the public lands for providing non-commodity resources, the RPA Programs placed a special emphasis on these programs. Each Program to date has recommended that commodity resource programs such as timber and minerals constitute a steadily declining share of overall funding for National Forest System management activities, and that non-commodity resource programs such as recreation, fish and wildlife habitat management, and soil and water protection should comprise a steadily increasing share of overall NFS funding. With the advent of the first RPA Program in 1975, the Forest Service made a fundamental change in its internal budget development process, moving from the traditional incremental budgeting model to one in which the budget targets in the RPA Program became the central focus in the development of budget alternatives.

Early experience using the RPA Program to guide the budget development and appropriations processes was encouraging. Throughout the late 1970s, the Forest Service pursued an aggressive budget strategy based on the RPA Program budget targets, at times requesting as much as double the previous year's appropriations for certain programs. This strategy seemed to pay off well for the Forest Service, with the NFS budget increasing nearly 30 percent in real terms between FY 1977 and FY 1980. The strategy seemed to pay off especially well for the non-commodity resource programs. Over the same period, funding for the recreation program rose over 40 per cent and for the fish and wildlife program more than 130 per cent. National Forest System budget growth was unprecedented and, in absolute terms, non-commodity programs grew more quickly than at any time in Forest Service history.

A broader and deeper examination of the events of this period raise serious questions, however, over how much of this change can be attributed to RPA planning. Although the average annual rate of increase in the budget for the National Forest System was more than three times that of the federal budget overall (net of defense and entitlement programs), it was only slightly more than that of the Bureau of Land Management, an agency in the Department of the Interior with land management responsibilities comparable to those of the Forest Service. Although it was experimenting at the time with a four-year budget authorization to overcome the usual single-year focus of the budget process, BLM did not have the benefit of any comprehensive planning process such as the RPA, and its appropriations came through the same congressional committees as those of the Forest Service. Over the 27-year period from FY 1960 to FY 1986, the average annual growth in BLM's appropriations was actually greater than that of the Forest Service.

Perhaps more importantly from a program planning viewpoint, congressional appropriations were apparently unswayed by the shift toward a greater emphasis on non-commodity resource programs recommended by each of the RPA Programs. Although appropriations for the non-commodity resource programs grew dramatically during the late 1970s, many of them continued to decline as a share of the overall NFS budget relative to the commodity resource programs.

This was quite the reverse of the budget emphasis recommended in the RPA Programs. The timber program share of NFS funding declined briefly after 1977 but soon began to climb again. Actual appropriations ran even more counter to the RPA Program recommendations in the early 1980s--resource programs continued to grow while many of the non-commodity resource programs showed significant real-term decreases. The Office of Management and Budget had advanced a 1980 RPA Program alternative, the so called "low-bound" alternative, which specified very different budget goals and program emphasis than did the Forest Service's "high-bound" alternative. OMB's alternative was summarily rejected and the Congress officially approved and adopted the Forest Service's high-bound alternative to guide national forest management, and the funding to carry out those activities, over the following five years.

In hindsight, however, a comparison of actual appropriations to the high-bound and low-bound budget targets show that Congress in effect did almost exactly what OMB had specified, not what the Forest Service had recommended--and Congress itself had endorsed--through RPA planning. By 1984, the Forest Service tacitly recognized the inefficacy of RPA as a budget tool. With the advent of the first RPA Program in the late 1970s, the Forest Service had directed its field staff to base its budget proposals on the RPA budget targets rather than on an incremental change from the previous year's budget. With the failure of the RPA budget targets to redirect congressional appropriations to the Forest Service, this approach was finally abandoned. Directions to the field for formulating the FY 1986 budget were to base the central budget alternative on the previous year's budget, leaving the RPA budget goals to define the most costly of the five budget alternatives, and the least likely to be funded. In essence, the Forest Service had returned to the classic incremental approach to budgeting, abandoning RPA as a means for effectively guiding the budget process toward the achievement of long-term resource management goals.

If RPA has indeed fallen short of its originators' expectations, it is important to understand why this is so, not only for the continued implementation of RPA but for the implementation of the forest plans prepared under NFMA. Some of the shortcomings of RPA planning are due to its failure to consider Forest Service programs in the context of a larger federal governmental system, and to consider the motivations and influence of other major organizational participants with that system. The overall incremental nature of decision making in large and complex organizational systems makes major shifts in program emphasis difficult to bring about quickly. The stability of the Forest Service's "iron triangle"--the agency itself, its congressional committees (both appropriations and standing) and the myriad national forest interest groups--serve to protect the Forest Service from change that might be perceived as having a negative impact on the agency, but it can also impose a sort of paralysis that prevents *any* significant change from the status quo, whether negative or positive.

The annual budget for the Forest Service is not so much the product of a unified, rational policy process as it is the product of organizational outputs and bureaucratic politics among the several different organizations that must act on the Forest Service's budget proposal. Each organization tends to play the same general role each year, based upon the history and values of each organization and the standard operating procedures that allow each organization to function effectively. Like any federal agency, the Forest Service tends to advocate increased expenditures for its programs. OMB acts as a presidential servant with a cutting bias, evaluating all budget proposals through the lens of "economy and efficiency" and relying heavily on marginal analysis as the standard determinant of whether a given program should be increased, cut or continued at all. The House, in this case the Interior appropriations subcommittee, is attuned to an agency's tendency to inflate its budget request and seeks to trim funding where necessary. Nonetheless, it is sensitive to and knowledgeable of the needs of the agency and often adds back funding for selected items singled out for cuts by OMB. The corresponding subcommittee in the Senate tends to act as an appeals court to which agencies and interest groups carry their disagreements with the House actions; the Senate often restores cuts made by the House. While the Senate pursues its own appropriations agenda, independent of the House and of the president, its positions often result in a compromise with the House in the conference committee, which produces an outcome not substantially different from that proposed in the President's Budget.

The complex of interests within the Congress, and the national forest interest groups that seek to influence its decisions, create a dense political environment in which significant policy change in any direction becomes exceedingly difficult to effect, especially in terms of budgets. An expanding budget "pie" relieves some of this external pressure, one reason the non-commodity resource programs did so well in the late 1970s, but it becomes increasingly difficult to effect a reallocation of resources when that pie is static or beginning to shrink. Congress is forced to identify its real priorities based on its own collective and individual needs. Members of Congress *must* be responsive to the needs of the local economies in their districts, and federal programs and policies have always been a key focus. Congress thus gives special attention in the appropriations process to federal projects or revenue sharing programs that are seen as supporting or promoting local economic development. In the case of the national forests, these sensitivities lead Congress to place a special emphasis on the timber sale program, not only because of the income and employment this generates through local industry but because local governments receive 25 percent of all national forest receipts, nearly all of which come from timber sales. In some western counties, such federal revenue sharing comprises more than 80 percent of the annual operating budget. To minimize local budget shortfalls and uncertainty, a clear message is sent to congressional representatives to maintain timber harvests at the highest level possible and to minimize the fluctuation in the harvest levels from one year to the next, economic conditions notwithstanding.

The stability of these organizational roles allows each participant to anticipate the actions of the other participants and develop a strategy designed to achieve a final outcome that it can find acceptable. This system can work smoothly until one of the participants steps out of role, inducing uncertainty and instability--both of which are actively avoided by most large organizations--in the relationships. More than any other participant in the Forest Service budget process, the actions of the assistant secretary of agriculture are determined not so much by organizational outputs as by bureaucratic politics and individual personality. In effect, this makes the Office of the Secretary the least predictable participant in the process, a wild card whose actions over time, and especially between presidential administrations, are much more variable than even those of OMB in the Executive Office of the President. The assistant secretaries who actually oversee the operations of the Forest Service have historically tended to be more caretakers than activists. But in recent years this has changed, partly due to a recent narrowing of the responsibilities of the office of the assistant secretary for natural resources and environment. Through steady, constant pressure, an activist assistant secretary can largely accomplish a personal budget agenda if he or she has a thorough understanding and respect for the organizational values, incentives and outputs of the Forest Service, OMB and the relevant committees in Congress. An activist assistant secretary who lacks this understanding and establishes adversarial relationships with other players in the budget process can be quickly isolated. The system then returns to the equilibrium defined by the organizational outputs of the major participants and the stable relationship between them.

The structure of the RPA planning, as determined by the legislation itself, is predisposed to instability vis-a-vis the assistant secretary by a schedule that is out of synchronization with the political calendar. The RPA Program is a plan nominally submitted to Congress by the secretary of agriculture. It is an opportunity for the incumbent assistant secretary to lay out his or her agenda for the Forest Service, and is thus an inherently political document. The 1980 RPA Program had been in preparation during the four years of the Carter administration and was being finalized as Ronald Reagan was preparing to assume the presidency. It should not be surprising, even if one discounts the differences in political philosophy between the two administrations, that Assistant Secretary Crowell was not particularly interested in implementing a plan negotiated and approved by his predecessor. With its five-year duration, it would have more than covered the entire period of the first Reagan administration.

Budget Execution

One finding from the case studies of budget development and execution on five national forests in FY 1986 is that even field budget proposals that have been developed on an incremental basis can get substantially rearranged during the budget process, resulting in difficulties in matching available funds to management tasks in the field. As field budget proposals become based on the budget requirements of the forest plans, many of which call for a shift in program emphasis not unlike that of the RPA Programs, such mismatches can be expected to occur more frequently and to a greater degree. Delays in the appropriations process pose additional difficulties for field managers. Final congressional action on Forest Service appropriations typically does not permit actual fund allocations until January, already more than three months into the fiscal year. A continuing conservative approach to fund allocation in some regions further compounds this problem for field managers. When substantial amounts of program funding are held at the regional offices to cover expenses there or simply as a means of more tightly controlling the distribution of funds among national forests throughout the fiscal year, the national forests and ranger districts do not know what funding will be available to them until literally the last weeks of the fiscal year. This makes it extremely difficult for field managers to accomplish well considered, integrated resource management and to use the funds efficiently.

An examination of both the funding and resource outputs that were proposed, allocated and reported for the five case study forests for FY 1986 shows no direct relationship between spending and output accomplishments for many of the resource programs. The general pattern among the studied forests is for the funding levels proposed by the field to be reduced during the budget process, but for there to be little or no reduction in the resource output levels also proposed by the field. In fact, for some resource programs, funding was reduced while the output targets were actually increased from the field budget proposal. Nonetheless, the accounting records and output target accomplishment reports generally show minimal departures from the assigned targets despite substantial reductions in funding. No doubt there are instances in which the field managers have inflated the estimated cost of accomplishing a given level of output. For the most part, however, field review on the case study forests indicates that, for many resource programs, either all but the most critical work (e.g., that necessary to protect the resources or previous capital investments) has been eliminated or deferred, or the work was done to a lower standard than was assumed in the original field budget proposal. These were the cases in which the reported expenditures and resource outputs were reasonably reliable; in most cases they were not. A combination of the softness in the estimates of many non-commodity resource outputs and propensity of field officials to allocate reported expenditures among resource programs on the basis of what was planned rather than what was actually worked raise serious concerns about the validity of any of the figures in the official record.

The Forest Service's strong organizational emphasis on meeting assigned resource output targets, and on containing issues at the ranger district level whenever possible, encourages field managers to "get the job done" first and worry about the accuracy of their accounting records secondarily. Through its periodic performance evaluations, which determine an individual's compensation and opportunities for professional advancement, the Forest Service has developed a powerful set of incentives focused primarily on meeting targets. Although the agency also strongly encourages its managers to maintain thorough and accurate records of all financial activities, this is not something that will materially influence a field official's performance evaluation (unless egregious errors or willful violations have occurred) nearly as much as the failure to meet his or her timber sale target. Through the Forest Service's well-established rewards system, an unspoken but unmistakable signal is sent to ranger district and national forest staff and line officers: Spend appropriations just as they are allocated and account for those expenditures accurately, but not at the cost of failing to meet resource output targets or letting some local issue get out of hand.

Part of the reason for this is that the agency itself no longer relies upon unit cost calculations flowing up from the districts in order to develop its own budget request. Since the adoption of Planning, Programming and Budgeting (PPB) by the Forest Service in the early 1970s, the accounting system is no more than a reporting requirement of the Congress to make sure that the Forest Service has not overspent its appropriations. Field managers have no strong disincentive against fudging the numbers somewhat to make reported expenditures match with allocated funding for each program account. Doing so has allowed them to accomplish their tasks despite the perceived fickleness of the politicians in Congress and, because it is no longer necessary to the Forest Service's internal budget development process, it has caused no operational difficulties to the organization itself. It can be argued that, until recently, no one in the Forest Service even knew the cumulative extent of the accounting irregularities; the agency had no need to be aware of such things as long as its records could show that it was within its congressional budget authority.

As a result of its initial investigations into below-cost timber sales, Congress now realizes that neither they nor the Forest Service really know what it costs to produce certain national forest outputs. The current accounting system is not explicitly designed to produce such information and what information is available is not reliable. The Forest Service has responded by issuing stern orders to field officials to "charge as worked," but agency leaders also realize that until the underlying incentive structure is changed, there will be problems with compliance that will continue to make the overall system untrustworthy. The Forest Service's second response, therefore, has been to propose to Congress that they control Forest Service activities not through a multitude of separate budget accounts that must each be tracked separately in the field, but through a more detailed set of output targets and broader appropriations that do not become obstacles to either accurate accounting or efficient, integrated resource management. GAO has pointed out that, although the proposed end-results budgeting system will ostensibly diminish the direct control Congress has over Forest Service activities by reducing the number of budget line items for which funding is specified, it would, in reality, provide a greater degree of control because the information Congress would receive would be more accurate than the information it now receives.[1]

There are still problems to overcome before the proposal for a different budget structure can make much more progress. Currently, the measurement of outputs for many of the resource programs is as unreliable, and in some cases virtually meaningless, and could not provide a sound basis for congressional action and oversight. There is a need for a much more clearly defined set of output targets that have a direct relationship to inputs of labor and funding. The targets should also provide some clear and objective indication of the *quality* of the outputs as well as the quantity; not only would these be more meaningful to members of Congress, they would facilitate the display of trends over time and permit Congress to respond before resource conditions declined to unacceptable levels. There is a further need to overcome a mutual mistrust between the Congress and the Forest Service field officials. Congress is miffed that it has been getting unreliable information all along, and is not inclined to provide broader discretion to the Forest Service until the agency proves it can account for its activities accurately. Field officials, on the other hand, are not inclined to provide more accurate unit costs until they have some assurance that the information will not be used to further reduce their flexibility to deal with their actual management responsibilities "on the ground." The chief of the Forest Service is in the key position to orchestrate this rapprochement through a gradual building of trust and confidence on both sides.

Forest Service Program Planning and the Legacy of PPB

RPA planning, as currently conceived and implemented, is a direct outgrowth of the Forest Service's adoption of the Planning, Programming and Budgeting system in the early 1970s. As such, it continues to exhibit many of the same characteristic shortcomings that caused PPB to be generally abandoned by federal agencies soon after it was introduced. PPB was conceived as a means of rationalizing an agency's budgeting process by (1) clearly defining the organization's goals and objectives, (2) developing a set of budget alternatives that directly linked funding levels to the achievement of those objectives, and (3) evaluating programs using reported accomplishments and expenditures and using this information in future program planning. It was felt that if the analysis of the alternatives was thorough enough, and if there was general agreement on the objectives, the agency's planning and analysis would be so compelling that budgets would naturally fall into line. Actual experience, however, showed that, as a tool for resource allocation and for shifting the funding emphasis to match the ostensible program emphasis, PPB was a failure. Agency after agency found that its budget process continued to be an incremental one, both as it was acted upon by the Congress and as it was developed internally, and that the eminently more rational budget alternatives flowing from the PPB approach had had a negligible impact on the allocation of budget resources. As this study's comparison of the RPA Program budget targets with actual appropriations over the past decade attests, this is an accurate description of the Forest Service's experience with RPA planning as well. Not only did the Congress continue to approach Forest Service budgeting in an incremental manner, even the agency request itself continued to reflect an incremental approach rather than the detailed annual budget goals established by the Forest Service--and ratified by Congress--in the RPA Programs. RPA's failure as a tool for shifting funding and program emphasis is further indicated in the fact that, to the extent this emphasis has shifted, it has shifted in the opposite direction of that called for in the RPA Programs.

Technocrats and born rational-comprehensive planners tend to blame the failure of RPA planning, like that of PPB generally, on the intransigence of self-serving politicians, but the structural problems identified in PPB seem to apply especially well to RPA and the Forest Service. PPB called for agencies and Congress to focus not on functional line items such as physical facilities or equipment that spanned several of an agency's programs, but on the programs and program outputs themselves. It has been argued, however, that such a focus does not make it easier to shift budgetary resources among programs but more difficult.

> Line-item budgeting, precisely because its categories (personnel, maintenance, supplies) do not relate directly to programs, are easier to change. Budgeting by programs, precisely because money flows to objectives, makes it more difficult to abandon objectives without abandoning the organization that gets its money for them.[2]

The reduction in a particular functional line item can easily be spread across many programs without threatening the organization's structure or specific groups of personnel within that structure. When a major reduction is proposed for a particular program, however, the automatic defense mechanisms within the agency itself and among that program's interest groups outside the agency are triggered. To the extent that the existing program emphasis is out of step with the desired program emphasis, it is more difficult for the agency to make changes under program budgeting than under functional line item budgeting. Thus, it is argued, program budgeting increases rather than decreases the cost of correcting error.[3] The upshot is that it may have been easier to effect a reallocation of budget resources among programs before the advent of PPB than after, a conclusion reached by many federal agencies that experimented with PPB.

It can be argued, however, that there has never actually been a "before-PPB" at the Forest Service, that its budgeting system has always had more in common with program budgeting than functional line item budgeting.[4] Long before the institution of PPB in the federal government generally, the Forest Service requested funding from Congress on the basis of National Forest System *programs*--the timber program, fish and wildlife program, recreation program, and so on. In fact, until 1982, even the functional costs associated with general administration and overhead were spread out by formula among these programmatic budget line items. The proliferation in budget line items over the past three decades has simply been a finer differentiation of programmatic funding (e.g., the timber program has been broken into timber sale preparation, harvest administration, timber resource planning and silvicultural examinations) not along functional lines. The Forest Service's programmatic budget structure has evolved in response to the nature of the agency's management responsibilities, and has for many years been supported by long range planning, first on the basis of individual programs and then merging these together into integrated, multiple-use resource management plans.

Such a structure poses obstacles for making changes that require significant reallocations of budgetary resources. This suggests that the difficulties encountered in implementing RPA planning may have originated as much within the Forest Service as with Congress, OMB or other participants in the agency's budget process. The fact that the Forest Service's own overall agency budget requests have continued to indicate an incremental approach and have not reflected the RPA budget targets any more than the OMB proposals or congressional appropriations reinforces this conclusion. The Forest Service fell into budgeting programmatically because the organization *thinks* programmatically and its managers, right down to the district rangers, *manage* programmatically. Asking such managers to also be planners and to propose alternatives that would involve major shifts in their responsibilities and *modus operandi* imposes significant internal tensions, and raises questions as to just how objectively the range of alternatives may have been formulated and evaluated. And the district rangers are not alone in their dilemma; the fact that each program is represented in four different hierarchical levels in an organization of 35,000 individuals suggests that bureaucratic needs for stability can even further constrain managers' ability to adopt any major changes in direction.[5]

One response to this inherent tension is to isolate planning from operational decision making and on-the-ground management:

> Organizations historically have dealt with the tension between planning and management by treating planning as a distinct and separate activity to be completed quickly and put on the shelf so that agency staff groups can "get back to business." Often a separate "planning team" is created, isolated from other staff units, and set off in a corner to do its work.[6]

RPA planning has always been done almost entirely by a small staff in the Washington office of the Forest Service. Organizationally, it is located under the deputy chief for programs and legislation, not under the deputy chief for the National Forest System. Though the RPA Program focuses almost entirely on national forest programs, the RPA planning staff is completely separate from the National Forest System branch. Field officials have little input into RPA planning beyond the transmitting of data to be compiled in the RPA Assessment, which sets no goals for the organization. Among the case study national forests and regional offices, officials interviewed did not feel the RPA Program had any particular relevance to the discharge of their daily responsibilities. By Washington office directive, the RPA goals are still the basis for the highest of the five budget alternatives in the field's annual budget submission; but this alternative is seldom used in developing the overall agency request. Also by Washington office directive, each forest plan must include an alternative based upon the RPA goals for that national forest; but most of the forest plans have now been completed and not one of them is based on an RPA alternative. This hardly suggests that there is a strong commitment to RPA planning among Forest Service field managers.

Implications for the Implementation of National Forest Planning

Although the National Forest Management Act was a legislative extension of RPA, and forest planning is ostensibly a part of the RPA planning process, it is not necessary that implementation of forest planning suffer the same shortcomings as RPA planning. In fact, in many ways, forest planning can be seen as the alter ego of

RPA planning. RPA planning is a centralized and centralizing process. It imposes top-down solutions to centrally identified problems, without significant input from either local managers or local publics. The benefits associated with achieving its specified goals, or more importantly the *costs* associated with *not* achieving those goals, are diffuse and difficult to relate to, for both the local publics and their members of Congress. In the political calculus of the Congress, there is no cost associated with ignoring the recommendations of the RPA Program. More than a decade after its establishment, few constituents are even aware it exists; thus, with no dearth of competing demands for their time and attention, fewer and fewer members of Congress need to know it exists. The RPA Program is directed to the Congress, yet many in Congress have so little regard for RPA planning that it has come close to being essentially terminated. For fiscal year 1987, the House and the Senate agreed to slash funding for RPA planning, and sent an unmistakable message to the Forest Service in the appropriations committee report:

> the Committee does not agree to continue to spend millions of dollars on planning documents that are not provided in a timely manner, do not reflect reasonable and professional judgments and estimates when they are released, and are not of particular value to the Committee when finally available.[7]

By sheer necessity, the forest planning process requires that Forest Service field managers play a substantial role in the planning process, if only because the resulting forest plans, unlike the RPA Program, are legally binding documents that could limit their future management flexibility. The tension between planning and management still exists for the field manager, raising the possibility of artificial limits being imposed on the scope and direction of the formulation, evaluation and selection of planning alternatives.[8] A distinguishing characteristic of NFMA, however, is that it specifies a planning process that is open at every stage to broad public participation, and there are many administrative and legal avenues for individuals and interests outside the Forest Service to see that a sufficiently broad range of alternatives is adequately evaluated. And this is not the only instance in which the public participation aspect of forest planning, though difficult and time-consuming for the agency, may be what saves it from the fate of RPA. Unlike RPA, forest planning is decentralized and decentralizing. In addition to providing essential bottom-up resource capability information, forest planning generally provides local solutions to locally identified problems, with those solutions often representing a negotiated agreement that has the tacit support of all the interested parties, including the Forest Service itself.[9] It is precisely because of this broad public role in developing the forest plans that members of Congress must take notice of them.

Again unlike RPA, the decisions made in forest planning have a direct impact on economic development and environmental quality that can often be clearly visible within the confines of a congressional district. No issue commands the attention of a member of Congress more effectively than a federal program that directly affects jobs and the quality of life for voters in his or her home district. If the forest plans represent agreements that have already garnered the broad based support of constituents, it is quite likely that they will also have the active support of members of Congress. Such support is most clearly expressed through annual Forest Service appropriations.

Another critically important factor to the successful implementation of forest planning is that public participation is no longer limited to the planning process, but extends, in effect, to the annual budget development process as well. In the budget development process that is now emerging with the completion of the forest plans, forest supervisors are being told to use the forest plan as the central focus of their budget alternatives. Typically, the forest supervisor is given a "bottom line" budget constraint at the outset of the budget development process, and he or she can then propose specific funding within that constraint on the basis of whatever mix of resource programs is judged to best implement the forest plan. Later, during the fund allocation process, the mix of funding that was specified becomes the most important determinant of how the actual program funds are distributed to that national forest. As the forest supervisor is developing budget alternatives, some assuming a slightly higher bottom line and some slightly lower, there is a great deal of flexibility for the forest supervisor to express his or her personal priorities among programs. As seen earlier, the optimal distribution of funding among the various resource programs may change with different levels of overall funding, and what constitutes "optimal" at any given funding level is within the discretion of the forest supervisor. For example, for resource programs given high priority, a wide range of projects will appear even at the lower overall funding level to be sure that these projects are covered in the event that subsequent fund allocations are lower than expected. In the progressively higher budget alternatives, more and more of the middle- and lower-priority projects in other resource programs will be included.

The critical question becomes: Which programs and projects are included, even in the lowest budget alternative, that will not get attention unless the subsequent fund allocation is substantially higher than expected? Each of the budget alternatives must be roughly within the spirit of the management emphasis represented in the forest plan. In anticipation of continued mismatches between planned and actual funding, the national forest regulations specify that the forest supervisor may alter the "implementation schedule," meaning that the portion of the planned resource management activities that receive adequate funding may be "implemented at a faster rate" than that portion not adequately funded.[10] A "contingency strategy" included in the plan for one of the study forests stated that

> If the forest plan is not fully funded during the next ten years then some scheduled management activities will have to be delayed, but the overall objectives stated in the plan will not change. . . . If we must accept budget cutbacks in particular program areas, such as timber roads, recreation or wildlife, then the least important activities in those areas will be delayed until adequate funding is received.[11]

This obviously raises important questions regarding the continued integrity of the plan as the blueprint for a particular balance of multiple use resource management, especially when one considers the compounding effect of several years of actual funding that is out of step with planned management activities. Under the regulations, the forest supervisor needs to give no public notice of shifts in management activities in response to funding constraints "unless the changes significantly alter the long-term relationship between levels of multiple-use goods and services projected under planned budget proposals as compared to those projected under actual appropriations." Action under this provision is not triggered until actual resource management activities are judged to have deviated so much from what was specified by the forest plan that an entirely new forest plan must be developed from scratch and the entire NEPA review process begun anew.[12]

However, there has been a recent interpretation of Forest Service budget development policy that, though subtle, could have a substantial impact on the role the public plays in the implementation, as well as the development, of national forest plans. As discussed above, OMB imposes tight restrictions on the public disclosure of agency budget proposals that are preliminary to the President's Budget. In the past, the Forest Service has interpreted this to apply to all budget development information assembled anywhere within the agency, from the field offices up to the chief. Public requests for information on what mix of funding a forest supervisor has proposed for a particular national forest were regularly denied on the basis of the OMB prohibition. In recent direction to all the field offices, however, the chief has interpreted this prohibition to apply only to the final Forest Service agency request developed at the Washington office. Budget proposal information from the field offices is regarded as staff recommendations meant only for internal agency use:

> Soliciting recommendations from field units on what should be done is not part of the budget development process. The Chief is the only line officer in the Forest Service who makes budget recommendations.[13]

The practical significance of this new interpretation is that the annual budget proposal for each national forest, which will later be used to guide the allocation of program funding to the forest, will now be open for public review and evaluation while there is still a measure of flexibility for making adjustments to the proposal. This will provide the forest supervisor an opportunity to show how well his or her funding priorities at each alternative budget level reflect the program balance represented in the forest plan:

> If requested by outside groups, Forest Supervisors should feel free to discuss what their priorities were within the alternative budget levels which display different rates of plan implementation.[14]

The key information in the budget proposal for many national forest users will be the mix of anticipated management activities at the lower budget alternatives and at the alternative that assumes essentially a continuation of the current overall funding level. While the forest supervisor may need to point out that maintaining the same program emphasis at several alternative budget levels does not necessarily imply the same funding distribution at each level, national forest users will be interested in seeing that there is adequate budget emphasis on each of the resource programs at the most likely funding levels as well as the higher budget alternatives.

Of course, with the many opportunities for modifications in the budget proposal, both within and outside the Forest Service, a field budget that is consistent with the forest plan is no guarantee of appropriations that are consistent with the forest plan. Under the current budget structure, the public gains only a greater certainty that there exists a budget document that can be used to link appropriations to planned programs to the extent possible. With the widening application of end-results budgeting, however, the balance of program funding in the field budget proposal is playing a much more substantial role. Within the constraint of the bottom line funding allocated to a given national forest, the forest supervisor will be more or less free to allocate the funding to individual resource programs according to his or her earlier budget proposal. At a time when field-level budget development is beginning to take on a much greater significance in the implementation of the forest plans, this process is being opened to the public, allowing individuals and organizations that have invested substantial effort in the completion of the plans to now follow through and help see their efforts made manifest in the actual management of the lands and resources of the national forests.

This policy refinement has benefits for local managers as well as for the public, though this was found to not be uniformly evident to officials on the case study forests. There is ambivalence among some officials who foresee the difficulties and complications of public involvement being introduced into an annual decision process with much firmer deadlines than were imposed on forest planning. Other officials, however, see this as an opportunity to enlist the public in their effort to see that plan implementation does not founder in the shoals of the budget process. Some forest supervisors have already produced a first annual report of their progress in implementing their forest plan, complete with charts comparing the program budget levels thought necessary for implementation to those actually received in appropriations. Although the forest supervisor cannot exhort local citizens to go forth and lobby Congress on behalf of the Forest Service budget, the implicit message in such information is plainly evident. More than this, however, if it is clear from the beginning that the forest supervisor's ability to manage the forest according to the "social contract" represented by the plan has been repeatedly limited by mismatched budgets, a local backlash against the Forest Service for noncompliance becomes less likely. The extent to which citizens can use the budget information to actually work to change the situation is simply a bonus for the local manager.

It is not inconceivable that this expanded public access to the Forest Service budget process could have substantial benefits for Congress as well. The public participation aspects of forest planning, combined with this opening to the budget development process, may have what is essentially a decentralizing and democratizing effect on the appropriations process. The final stages of the budget and appropriations process are arcane and complicated and are conducted largely in the committee rooms and private offices of the Congress, thousands of miles from most national forest users. The sheer volume of budget detail that even the Interior appropriations subcommittees must deal with (funding more than a dozen agencies in at least four different cabinet departments) necessitates a reliance on interest groups and lobbyists to help identify issues and develop alternatives. Because of the complexity and centralization of this process, Washington-based interest groups and lobbyists are often the *only* ones outside the government itself with any effective access to it. A set of forest plans representing a rough consensus on how the national forests should be managed becomes another source of information for congressional decision makers, indirectly giving voice to thousands of national forest users at the local level. There may be more direct local participation in the appropriations process as well. Involvement in forest planning, and now in budget development, can educate local users on the intricacies of agency and congressional processes and increase the number of people who seek direct involvement with their legislators regarding the budgets for national forest management. To the extent that this too becomes an additional source of information for congressional decision makers, there is correspondingly less reliance on the centralized interest groups and lobbyists and a broadening of the base of public input into what was heretofore a relatively closed process.

Toward A True Integration of Planning, Budgeting and Management

As an organization, the Forest Service has long quested for a single system to unify its three key management processes: long-range resource management planning, annual budgeting and the management of day-to-day operations on the national forests. Due to a variety of fundamental structural and political faults, the three processes are still out of step with one another and tend to pull the agency in three different directions at once. The Forest Service's sheer organizational will to make these processes work together has only served to compromise the validity of administrative recordkeeping, from the ranger districts on up, so that even the Forest Service leaders themselves no longer have the means to reliably assess the extent of the integration problem. Clearly, the time has come for the agency to rethink some of its traditional approaches to these three processes and how they are linked together.

If persistent problems with the definition of resource outputs can be overcome, the Forest Service proposal for end-results budgeting may have greater potential for process integration than even the agency itself recognizes. A *quid pro quo* in which Forest Service field managers dedicate themselves to more accurate reporting of both resource outputs and program expenditures in return for greater flexibility to spend National Forest System appropriations in whatever manner best addresses the local management situation, will go far toward better coordinating budgeting with day-to-day operations. In effect, this will permit the Forest Service to finally skirt one of the major pitfalls identified in the PPB system. It has been shown that budgeting by program inevitably triggers bureaucratic defense mechanisms and makes it extremely unlikely that any significant reallocation of budgetary resources can occur, even if a shift in program emphasis was called for in program plans developed by the agency itself. Taking appropriations control above the level of the individual resource programs, as in the end-results budgeting proposal, has an effect much the same as going to a functional budget line item structure. An appropriation can be increased or decreased without being perceived, either by the agency or its interest groups, as a direct threat to any particular program. The appropriations process can continue, as it apparently must, to effect only incremental change in the agency budget. But if the legal and bureaucratic obstacles to reallocating resources *within* that appropriation are removed, meaningful change at the program level becomes possible.

But this is only half of the equation. The end-results approach will make budgeting more compatible with planning but, if a successful integration is to be achieved, planning must also seek to become more compatible with budgeting--the budgeters must think somewhat more like planners and the planners must think somewhat more like budgeters. The end-results approach makes budgeting more like planning in that it focuses on a thorough harmonization of resource programs on the ground instead of on the compartmentalization of expenditures in numerous, detailed accounts. It broadens the flexibility of field managers to respond to demands or management situations that were not foreseen by the managers themselves during budget development or by Congress during the appropriations process. But this gesture must be met halfway by making planning more like budgeting. To accomplish this, planners must adopt a realistic view of the relevant range of proposals (i.e. what might possibly be approved) and then generate a set of alternative proposals that covers the relevant range.

Proponents of the PPB system have noted that "as long as the budget allocation which comes down is not too different from the program proposals then PPB can work."[15] But this observation is valid in the reverse as well. If even a modified PPB approach is to work, the program proposals cannot be too different from the budget allocation that can reasonably be expected. RPA Programs and forest plans cannot continue to be based on the assumption that the agency budget will double within the next five years. The prevailing notion that program plans with stratospheric budgets must be produced so that Congress will be swayed by the overwhelming opportunity costs associated with *not* providing full funding rings rather hollow considering the Forest Service experience with RPA. It is more in the interest of the Forest Service itself than of the Congress that the agency get its planning and budgeting systems coordinated. Congress will continue to function in any case; long range planning at the Forest Service may not.

There are two measures that can be taken by the Forest Service to better integrate planning with the realities of the budget process. First, planning alternatives could be based on assumptions of overall funding not to exceed, say, 110 percent of current funding in constant dollars. Different management emphasis in each of the planning alternatives might assume different mixes of funding among resource programs (timber, recreation, wildlife, range, soil and water), but keeping the "bottom line" within a reasonable range of the current funding trend would provide greater certainty that, at least from a budget perspective, any one of the alternatives could actually be implemented. This approach was taken in one Forest Service region in the development of its first set of forest plans. Forest supervisors in this region subsequently found themselves with sensible, implementable forest plans, while in other regions forest supervisors were already going back to their local publics to explain why they couldn't do what they said they would do (and generally laying the blame on Congress). The plans serve a key role in the allocation of funds by the chief's office to the regions, however, and basing its plans on reasonable budget assumptions has hurt the one region in this internal administrative competition. Thus, for this approach to work it would have to be adopted by all regions to reestablish a level playing field internally in the agency. The Forest Service is notably tradition-bound and making such a change could well require some prompting from the Congress.

Second, and perhaps as an alternative, the plans could display the resource managers' estimate of the optimal distribution of funding among the resource programs at a variety of overall funding levels less than 100 percent. If only 70 percent of the ideal funding level is available, it may not make sense to reduce all programs by 30 percent across the board. This mechanistic Gramm-Rudman approach to budget reduction takes no account of priorities among programs, nor does it recognize the interactions among resource programs. At 100 percent funding, there may be many "nice to do" projects that would not appear at all at 70 percent funding; on the other hand, key programs or projects may receive the same level of commitment at the lower level as at the higher. Displaying such information in the forest plan would get this message across to national forest users at the outset, conveying the managers' best estimate of how the thrust and emphasis of the forest plan would be maintained, even under reduced budgets. Early disclosure may also help avoid ill will when, during implementation, the various boats seem to be settling unequally as the budgetary tide recedes.

This second measure could also be a useful tool for Congressional budgetary decision makers. A primary purpose of the plans is to establish, with the help of broad-based public involvement, what the management emphasis should be on the National Forest System, for individual national forests and for the system as a whole. Plans that show the relative emphasis among resource programs only at an unattainable 100 percent funding level are of little guidance to Congress when the difficult choices must be made on how to best distribute a much more limited budget. Using the integrated resource management approach called for in NFMA, and relying on an understanding of resource interactions that is far greater than that of most members of Congress, Forest Service resource professionals can communicate to Congress as well as to the public how funds can best be allocated among resource programs at reduced budget levels and still maintain the management emphasis established in the plans. Congress needs little guidance on spending when their budget is unlimited; agency guidance on resource protection and management priorities is most critical when the budget is severely constrained.

The Forest Service estimates the cost of producing this first generation of forest plans for all the national forests to be more than a half billion dollars. The draft and final plans and environmental impact statements, even without the background files and documentation, comprise more than a quarter million pages and fill more than 60 feet of bookshelf. They have diverted the energies of thousands of professional resource managers and citizens for the more than ten years they have been under development. Whether this staggering investment results in better management of our national forests or produces little more than a very large and very expensive stack of paper will be determined by the willingness of the Forest Service, and the Congress, to innovate. The Forest Service planning and budgeting approaches of the past have not been well integrated and, if Congress' growing skepticism and neglect for RPA is any indication, the contributions of past planning efforts to improve resource management have fallen short of justifying their costs.

This study was conducted to critically and objectively examine the nature of these shortcomings, to better explain these past events and reveal the elements critical to policy and program failure. It is hoped that, in so doing, the elements critical to policy and program success are revealed as well. In 1974, Senator Humphrey and others began a grand experiment to determine whether long-range program planning, an intuitively sensible approach to the management of the nation's renewable natural resources, could actually be made to work in the political context of federal budgeting. We now have fifteen years of experience in which, like any worthwhile experiment, some things have worked as expected and others have not. The wrong turns that were made can be as valuable as those that were right, as long as they are recognized as lessons by which to guide future planning efforts. The management of our renewable natural resources on the basis of publicly affirmed long-term goals and objectives is still an eminently sensible idea. Through achieving the more effective integration of resource management planning, annual budgeting and local operational decision making, the keys to making this idea succeed in the context of national forest planning are within our grasp.

NOTES

1. U.S. Comptroller General, *Forest Service: Evaluation of "End Results" Budgeting Test*, p. 6.

2. Wildavsky, The *Politics of the Budgetary Process*, p. 201.

3. Ibid., p. 200.

4. The terminology becomes confusing here. What the planning and budgeting literature (and this study) refer to as "programs," Forest Service personnel often refer to as "functions." For example, the Forest Service has long referred to plans focused on a particular program, such as the timber program, as a "functional" plan; staffs for a particular program are typically referred to as "functional staff" rather than "program staff." This is perhaps because the larger divisions of the Forest Service (Research, National Forest System, State & Private Forestry) have come to be referred to as "programs" and another term was needed to further differentiate

within these divisions.

5. Kaufman, Herbert, *The Limits of Organizational Change* (University, AL: University of Alabama Press, 1971). See also: Downs, Anthony, *Inside Bureaucracy* (Boston: Little, Brown, 1976).

6. Cortner, Hanna J. and Schweitzer, Dennis L., "Institutional Limits to National Public Planning for Forest Resources: The Resources Planning Act," *Natural Resources Journal*, Vol. 21, No. 4 (April 1981), pp. 203-222 at p. 213.

7. U.S. Congress, Senate, Committee on Appropriations, *Department of Interior and Related Agencies Appropriation Bill, 1987, Report*, Senate Report No. 99-397, August 11, 1986 (Washington, D.C.: Government Printing Office, 1986).

8. Cortner and Schweitzer, "Institutional Limits to National Public Planning for Forest Resources," p. 214.

9. Collins, Richard C., "Resolving Forest Management Planning Disputes: A Mediator's Perspective," *Resolve*, Vol. 20 (Summer, 1988). See also: Wondolleck, Julie M., *Public Lands Conflict and Resolution: Managing National Forest Disputes* (New York: Plenum, 1988).

10. 36 CFR 219.11(e).

11. USDA Forest Service, *Green Mountain and Finger Lakes National Forest, Final Land and Resource Management Plan*, (Rutland, VT: USDA Forest Service, 1986), p. 5.05.

12. Ibid.

13. Letter from F. Dale Robertson, Chief, Forest Service to Regional Foresters, Station Directors and Area Director, "Forest Program Budget Alternatives," January 22, 1988.

14. Ibid.

15. Gremillion, McKenney, and Pyburn, "Program Planning in the National Forest System," p. 230.

Bibliography

Allison, Graham T., "Conceptual Modes and the Cuban Missile Crisis," *American Political Science Review*, Vol. 53 (1969) pp. 689-718.

Allison, Graham T., *Essence of Decision: Explaining the Cuban Missile Crisis*, (Boston: Little, Brown and Company, 1971).

Bachrach, Peter, and Baratz, Morton S., "Decisions and Non-decisions: An Analytical Framework," *American Political Science Review*, Vol. 57 (1963) pp. 632-642.

Bachrach, Peter, and Baratz, Morton S., "The Two Faces of Power," *American Political Science Review*, Vol. 56 (1962) pp. 947-952.

Bailey, John J., and Robert J. O'Connor, "Operationalizing Incrementalism: Measuring the Muddles," *Public Administration Review*, Vol. 35, No. 1 (January/February 1975), pp. 60-66.

Bardach, Eugene, *The Implementation Game: What Happens After A Bill Becomes Law*, (Cambridge, MA: MIT Press, 1977).

Behan, Richard W., "RPA/NFMA: Time to Punt," *Journal of Forestry*, Vol. 79 (1981).

Berman, L., *The Office of Management and Budget and the Presidency, 1921-1979*, (Princeton, NJ: Princeton University Press, 1979)

Bierman, Harold, *Implementing Capital Budgeting Techniques*, (Cambridge, MA: Ballinger, 1988).

Binkley, C.S., Brewer, G.D. and V.A. Sample (eds.), *Redirecting the RPA: Proceedings of the 1987 Airlie House Symposium on the Resources Planning Act*, Bulletin No. 95, (New Haven: Yale School of Forestry & Environmental Studies, 1988).

Blau, Peter M. and Marshall W. Meyer, *Bureaucracy in Modern Society*, (New York: Random House, 1956).

Brauer, Carl, "Tenure, Turnover, and Postgovernment Employment Trends of Presidential Appointees," in: Mackenzie, G. Calvin (ed.), *The In-and-Outers: Presidential Appointees and Transient Government in Washington* (Baltimore: Johns Hopkins University Press, 1987).

Collins, Richard C., "Resolving Forest Management Planning Disputes: A Mediator's Perspective," *Resolve*, Vol. 20 (Summer, 1988).

Congressional Budget and Impoundment Control Act of 1974, (P.L. 93-344).

Cooper, John M., "Gifford Pinchot Creates a Forest Service," in: Doig, Jameson W. and Erwin C. Hargrove (eds.), *Leadership and Innovation: A Biographical Perspective on Entrepreneurs in Government*, (Baltimore: Johns Hopkins University Press, 1987).

Corson, John J. and R. Paul Shale, *Men Near the Top: Filling Key Posts in the Federal Service*, (Baltimore: Johns Hopkins Press, 1966).

Cortner, Hanna J. and Dennis L. Schweitzer, "Institutional Limits to National Public Planning for Forest Resources: The Resources Planning Act," *Natural Resources Journal*, Vol. 21, No. 4 (April 1981), pp. 203-222.

Council of Economic Advisors, *Economic Report of the President*, Fiscal Year 1987 (Washington, DC: Government Printing Office, 1988).

Crecine, John P., "A Computer Simulation Model of Municipal Budgeting," *Management Science*, Vol. 13, No. 7 (1967), pp. 786-815.

Crecine, John P. and Gregory Fisher, *On Resource Allocation in the Department of Defense*, (Ann Arbor, MI: Institute for Public Policy Studies, 1971).

Crowell, John B., "Would I Do It Again?" Speech before the Western Timber Association, Portland, Oregon, May 1985.

Dahl, Robert, *A Preface to Democratic Theory*, (Chicago: University of Chicago Press, 1956).

Davis, Otto A., Dempster, M.A.H. and Aaron Wildavsky, "A Theory of the Budgetary Process," *American Political Science Review*, Vol. 60, No. 3 (1966), pp. 529-547.

Davis, Otto, M.A. Dempster and Aaron Wildavsky, "A Theory of the Budgetary Process," *American Political Science Review*, Vol. 40, No. 3 (1966), pp. 529-547.

Dawes, Charles G., *The First Year of the Budget of the United States*, (New York: Harper Brothers, 1923.).

deBuys, William, *Enchantment and Exploitation: The Life and Times of a New Mexico Mountain Range* (Albuquerque, NM: University of New Mexico Press, 1985).

Downs, Anthony, *Inside Bureaucracy*, (Boston: Little, Brown, 1976).

Dunsire, Andrew, *Implementation in a Bureaucracy*, (New York: St. Martin, 1979).

Etzioni-Halevy, Eva, *Bureaucracy and Democracy: A Political Dilemma*, (Boston: Routledge and Kegan Paul, 1983).

Executive Office of the President, Bureau of the Budget, *Staff Orientation Manual*, 1945.

Executive Office of the President, Office of Management and Budget, *Instructions on Budget Execution*, Circular A-34, Part III (revised August 26, 1985).

Executive Office of the President, Office of Management and Budget, *Preparation and Submission of Budget Estimates*, Circular A-11, Revised June 8, 1987.

Federal Anti-Deficiency Act, 31 U.S.C. 1351 and 1517. See also: Executive Office of the President, Office of Management and Budget, *Instructions on Budget Execution*, Circular A-34, Part III (revised August 26, 1985).

Federal Timber Contract Payment Modification Act of 1984, (P.L. 98-478).

Fedkiw, John, "The USDA Decision Process for the 1980 RPA," in: Hewett, Charles and Thomas Hamilton, *Forests in Demand: Conflicts and Solutions*, (Boston: Auburn House, 1982).

Fenno, Richard, "The House Appropriations Committee as a Political System: The Problem of Integration," *American Political Science Review*, Vol. 56, No. 2 (1962).

Fenno, Richard F., *The Power of the Purse: Appropriations Politics in Congress*, (Boston: Little, Brown and Co., 1966).

Fesler, James W., "The Basic Theoretical Question: How to Relate Area and Function," in: Grosenick, Leigh E. (ed.), *The Administration of the New Federalism: Objectives and Issues*, (Washington, DC: American Society for Public Administration, 1973).

Fesler, James W., *Area and Adminstration*, (University, AL: University of Alabama Press, 1949).

Forest and Rangeland Renewable Resources Planning Act of August 17, 1974, (P.L. 93-378, 88 Stat. 476, as amended; 16 U.S.C. 1601(note), 1600-1614).

Gremillion, Lee L., McKenney, James L. and Philip J. Pyburn, "Program Planning in the National Forest System," *Public Administration Review*, Vol. 40, No. 3 (May/June 1980), pp. 226-230.

Halperin, Morton, *Bureaucratic Politics and Foreign Policy*, (Washington, DC: Brookings Institution, 1974).

Harlow, Robert L., "The Decline and Possible Fall of PPB," *Public Finance Quarterly*, (January 1972), pp. 85-115.

Harper, Edwin L., Kramer, Fred A. and Andrew M. Rouse, "Implementation and the Use of PPB in Sixteen Federal Agencies," *Public Administration Review*, Vol. 29, No. 6 (Nov/Dec 1969), pp. 631-632.

Heclo, Hugh, *A Government of Strangers: Executive Politics in Washington*, (Washington, DC: Brookings Institution, 1977).

Hill, Jeffrey, "Why So Much Stability: The Role of Agency Determined Stability," *Public Choice*, 46(3):275-287.

Hilsman, Roger, *To Move A Nation*, (New York: 1967), pp. 554-555.

Humphrey, Hubert H., *The Education of a Public Man*, (Garden City, NY: Doubleday, 1976).

Ickes, Harold L., *The Secret Diaries of Harold Ickes: The First Thousand Days*, (New York: Simon and Schuster, 1953)].

Katz, Daniel and Robert Kahn, *The Social Psychology of Organizations*, (New York: Wiley, 1966).

Kaufman, Herbert, *The Forest Ranger: A Study in Administrative Behavior*, (Baltimore: Johns Hopkins University Press for Resources for the Future, 1967).

Kaufman, Herbert, *The Limits of Organizational Change*, (University, AL: University of Alabama Press, 1971).

Knutson-Vandenberg Act of June 9, 1930, 46 Stat. 527, as amended; 16 U.S.C. 576-576b.

Leman, Christopher K., "Planning Against Analysis: Forest Service Implementation of the Resources Planning Act," in: Binkley, C.S., Brewer, G.D. and V.A. Sample (eds.), *Redirecting the RPA: Proceedings of the 1987 Airlie House Symposium on the Resources Planning Act*, Bulletin No. 95, (New Haven: Yale School of Forestry & Environmental Studies, 1988).

Leman, Christopher K., *Resource Assessment and Program Development: An Evaluation of Forest Service Experience Under the Resources Planning Act, with Lessons for Other Natural Resources Agencies*, (Unpublished Paper), (Washington, DC: U.S. Department of the Interior, Office of Policy Analysis, August 1980).

LeMaster, Dennis C., "Federal Budgeting, Gramm-Rudman-Hollings, and Natural Resource Program Funding," in: *Proceedings of the Society of American Foresters Annual Meeting, 1986*, (Bethesda, MD: Society of American Foresters, 1987).

LeMaster, Dennis C., *Decade of Change: The Remaking of Forest Service Statutory Authority dur the 1970s*, (Westport, CT: Greenwood Press, 1984).

Lewis, Eugene, *Public Entrepreneurship: Toward a Theory of Bureaucratic Political Power*, (Bloomington, IN: Indiana University Press, 1984).

Lindblom, C.E., "Still Muddling, Not Yet Through," *Public Administration Review*, Vol. 39, 1979.

Lindblom, C.E., "The Science of Muddling Through," *Public Administration Review*, Vol. 19, Spring 1959.

Lindblom, C.E. and D. Braybrooke, *A Strategy of Decision*, (Glencoe, IL: 1963).

Lindblom, Charles, E., *The Intelligence of Democracy*, (New York: Free Press, 1965).

Lorange, Peter, *Implementation of Strategic Planning*, (Englewood Cliffs: Prentice-Hall, 1982).

Mackenzie, G. Calvin (ed.), *The In-and-Outers: Presidential Appointees and Transient Government in Washington*, (Baltimore: Johns Hopkins University Press, 1987).

Macy, John W., Adams, Bruce, Walter, J. Jackson and G. Calvin Mackenzie, *America's Unelected Government*, (Cambridge, MA: Ballinger, 1983).

Mazmanian, Daniel A. and Paul A. Sebatier, *Implementation and Public Policy*, (Glenview, IL: Scott, Foresman, 1983).

McKenney, J. and E. Cale, *An Analysis of the 1979 PPB Process of the U.S. Forest Service*, Unpublished report to the U.S. Department of Agriculture, Harvard Graduate School of Business Administration, Soldier's Field, MA, November 1977.

Merton, Robert K., "Bureaucratic Structure and Personality," in: Merton, R.K., Gray, Ailsa P., Hockey, Barbara and Hanon Selvin (eds.), *Reader in Bureaucracy*, (Glencoe, IL: Free Press, 1952).

Mouzelis, Nicos, *Organization and Bureaucracy*, (Chicago: Aldine, 1967).

Natchez, Peter B. and Irvin C. Bupp, "Policy and Priority in the Budgetary Process," *American Political Science Review*, Vol. 67, No. 12 (1973), pp. 951-963.

National Forest Management Act of October 22, 1976; P.L. 94-588, 90 Stat. 2949, as amended; 16 U.S.C. 472a, 476, 500, 513-516, 518, 521b, 528(note), 576b, 594-2(note), 1600(note), 1601(note), 1600-1602, 1604, 1606, 1608-1614).

National Forest Roads and Trails Act of October 13, 1964, (78 Stat. 1089, as amended; 16 U.S.C. 532-538).

National Forest System Land and Resource Management Planning, in: Code of Federal Regulations, Title 36, Part 219 (Washington, DC: General Services Administration, Office of the Federal Register, July, 1984), pp. 45-71.

Naylor, T.H., *Strategic Planning Management*, (Oxford, Ohio: Planning Executives Institute, 1980).

Neustadt, Richard E., *Presidential Power: The Politics of Leadership*, (New York: Wiley and Sons, 1976).

Novick, David, *Program Budgeting*, (Cambridge, MA: Harvard University Press, 1967).

Padgett, John F., "Bounded Rationality in Budgetary Research," *American Political Science Review*, Vol. 74, No. 4 (1980), pp. 354-372.

Parris, Judith, *The Office of Management and Budget: Background, Responsibilities, Recent Issues*, Congressional Research Service Report 78-158, (Washington, DC: Library of Congress, 1978).

Pastor, Robert, "Continuity and Change in U.S. Foreign Policy: Carter and Reagan on El Salvador," *Journal of Policy Analysis and Management*, 3(2): 175-176, (1984).

President's Commission on Economy and Efficiency, *Message of the President Submitting for the Consideration of Congress A Budget*, Senate Document 1113, (Washington, DC: Government Printing Office, 1913).

Pressman, Jeffrey L. and Aaron Wildavsky, *Implementation: How Great Expectations in Washington Are Dashed in Oakland; Or, Why It's Amazing That Federal Programs Work At All, This Being a Saga of the Economic Development Administration As Told by Two Sympathetic Observers Who Seek to Build Morals on a Foundation of Ruined Hopes*, Third Edition, (Berkeley: University of California Press, 1984).

Price, Donald K., "General Dawes and Executive Staff Work," *Public Administration Review* 10 (2) 1941.

Ripley, Randall B., *Congress: Process and Policy*, Second Edition (New York: W.W. Norton, 1978).

Robinson, Glen O., *The Forest Service: A Study in Public Land Management*, (Baltimore: Johns Hopkins University Press for Resources for the Future, 1975).

Sample, V. Alaric, "Improving the Linkage Between the RPA Program and the RPA Assessment Findings: The View From the Office of Management and Budget," in: Binkley, C.S., Brewer, G.D. and V.A. Sample (eds.), *Redirecting the RPA: Proceedings of the 1987 Airlie House Symposium on the Resources Planning Act*, Bulletin No. 95, (New Haven: Yale School of Forestry & Environmental Studies, 1988).

Sample, V. Alaric, "The Resources Planning Act (RPA) Process and Its Implementation to Date," in: Binkley, C.S., Brewer, G.D. and V.A. Sample (eds.), *Redirecting the RPA: Proceedings of the 1987 Airlie House Symposium on the Resources Planning Act*, Bulletin No. 95, (New Haven: Yale School of Forestry & Environmental Studies, 1988).

Sample, V. Alaric, "The Rise and Fall of the Resources Planning Act: 1974-1985," unpublished paper, Yale University, School of Organization and Management, New Haven, CT, 1986.

Sample, V. Alaric, and Peter C. Kirby, *National Forest Planning: A Conservationist's Guide*, (Washington, DC: The Wilderness Society, 1983).

Schick, Allen, "A Death in the Bureaucracy: The Demise of Federal PPB," *Public Administration Review*, Vol. 33, No. 2 (March/April 1973), pp. 146-156.

Schick, Allen, *Congress and Money: Budget, Spending and Taxing*, (Washington, DC: The Urban Institute, 1980).

Shands, W. and R. Healy, *The Lands Nobody Wanted: Policy for the National Forests of the Eastern United States*, (Washington, DC: The Conservation Foundation, 1977).

Shuman, H.E., *Politics and the Budget: The Struggle Between the President and the Congress*, (Englewood Cliffs, NJ: Prentice-Hall, 1984).

Sorenson, Theodore, "You Get to Walk to Work," *New York Times Magazine*, March 19, 1967.

Stonich, Paul J. (ed.), *Implementing Strategy: Making Strategy Happen*, (Cambridge, MA: Ballinger, 1982).

Stuckenbruck, Linn C. (ed.), *Implementation of Project Management: The Professional's Handbook*, (Reading, MA: Addison-Wesley, 1981).

Sundquist, James L., *The Decline and Resurgence of Congress*, (Washington, DC: The Brookings Institution, 1981).

Thompson, James C., *Rolling Thunder: Understanding Policy and Program Failure*, (Chapel Hill, NC: University of North Carolina Press, 1980).

Truman, David B., *The Governmental Process*, (New York: Knopf, 1961).

Twenty-five Percent Fund Act of May 23, 1908, (35 Stat. 259, 260, 267 (1908) as amended; 16 U.S.C. 500, 671; 31 U.S.C. 534).

U.S. Commission on Economy and Efficiency, *The Need for a National Budget*, House Document 854, (Washington, DC: Government Printing Office, 1912).

U.S. Comptroller General, *Congress Needs Better Information on Forest Service's Below-Cost Timber Sales,* GAO-RCED-84-96, 6/28/84, (Washington, DC: General Accounting Office, 1984).

U.S. Comptroller General, *Forest Service: Evaluation of "End-Results" Budgeting Test*, Report to the Subcommittee on Interior and Related Agencies, Committee on Appropriations, House of Representatives, GAO/AFMD-88-45, (Washington, DC: General Accounting Office, 1988).

U.S. Comptroller General, *Principles of Federal Appropriation Law*, (Washington, DC: General Accounting Office, 1986).

U.S. Comptroller General, *Timber Sale Accounting: Analysis of the Kootenai National Forest,* GAO/AFMD-86-62BR, 6/6/86, (Washington, DC: General Accounting Office, 1986).

U.S. Comptroller General, *Timber Sale Accounting: Analysis of Forest Service's Proposed Timber Program Information Reporting System*, GAO/AFMD-86-42, 4/4/86, (Washington, DC: General Accounting Office, 1986).

U.S. Congress, House of Representatives, Committee on Appropriations, *Hearings on Appropriations for the Department of Interior and Related Agencies for Fiscal Year 1989*, (Washington, DC: Government Printing Office, 1988).

U.S. Congress, House of Representatives, Committee on Appropriations, *Hearings on Appropriations for the Department of Interior and Related Agencies for Fiscal Year 1985*, (Washington, DC: Government Printing Office, 1985).

U.S. Congress, House of Representatives, Committee on Appropriations, *House Joint Resolution 465, Further Continuing Appropriations for Fiscal Year 1986,* House Report No. 99-450, (Washington, DC: USGPO, 1986).

U.S. Congress, Senate, Committee on Appropriations, *Department of Interior and Related Agencies Appropriation Bill, 1987, Report*, Senate Report No. 99-397, August 11, 1986, (Washington, DC: Government Printing Office, 1986).

U.S. Department of Agriculture, *Budget Estimates for the U.S. Department of Agriculture for the Fiscal Year Ending September 30, 1986*, (Washington, DC: Government Printing Office, 1985).

U.S. Senate, Committee on Agriculture, Nutrition and Forestry, *Compilation of the Forest and Rangeland Renewable Resources Planning Act of 1974,* (Washington, DC: Government Printing Office, 1979).

U.S. Senate, Committee on Appropriations, *Appropriations for the Department of the Interior and Related Agencies for Fiscal Year 1986*, Senate Report 99-141, (Washington, DC: USGPO, September 24, 1985).

USDA Forest Service, *Forest Service Manual*, Title 1930.2, as amended September, 1985, (Washington, DC: USDA Forest Service, 1985).

USDA Forest Service, *Green Mountain and Finger Lakes National Forest, Final Land and Resource Management Plan*, (Rutland, VT: USDA Forest Service, 1986).

USDA Forest Service, *Program Development and Budgeting Handbook*, Forest Service Handbook 1909.13.

USDA Forest Service, *Report of the Forest Service for Fiscal Year 1983*, (Washington, DC: USGPO, 1984).

USDA Forest Service, *Report of the Forest Service, Fiscal Year 1985*, (Washington, DC: USDA Forest Service, 1986).

Weber, Max, *The Theory of Social and Economic Organization*, translated by A.M. Henderson and Talcott Parsons, (New York: Oxford University Press, 1947).

Wildavsky, Aaron, *Budgeting: A Comparative Theory of Budgetary Processes*, (New Brunswick, NJ: Transaction Books, 1985).

Wildavsky, Aaron, *The New Politics of the Budgetary Process*, (Glenview, IL: Scott, Foresman, 1988).

Wildavsky, Aaron, *The Politics of the Budgetary Process*, Third Edition, (Boston: Little, Brown and Company, 1979).

Wilkinson, Charles F. and H. Michael Anderson, *Land and Resource Planning in the National Forests*, (Washington, DC: Island Press, 1987).

Willoughby, W.F., *The Problem of a National Budget*, (New York: Appleton, 1918).

Wolf, Robert E., "Architects and Architecture," *The RPA Process: Moving Along the Learning Curve*, Gerald E. Stairs and Thomas E. Hamilton, eds., (Durham, NC: Duke University Press, 1983).

Wondolleck, Julie M., *Public Lands Conflict and Resolution: Managing National Forest Disputes*, (New York: Plenum, 1988).

Zimmerman, Kristen E., *Capital Budgeting and Its Pertinence to the Management of the U.S. Forest Service*, Unpublished Research Paper, Harvard Graduate School of Business Administration, January, 1976.

Index

About the Author

V. ALARIC SAMPLE is a Senior Fellow at the Conservation Foundation in Washington, D.C.